Desertification

Other Titles in This Series

Westview Special Studies in Natural Resources and Energy Management

Desertification:
Environmental Degradation in and around Arid Lands
edited by Michael H. Glantz

The United Nations Water Conference (in Argentina in March 1977) and the United Nations Conference to Combat Desertification (in Kenya in August 1977) reflect the worldwide attention that recent global food shortages and growing populations have drawn to the destruction of arable and potentially arable land. This collection of articles focuses on a primary form of such destruction: desertification—the creation of desert-like conditions in arid or semiarid regions either by changes in climate patterns or by human mismanagement, or both. The contributors—representing a range of disciplines—examine and evaluate the social, political, economic, environmental, and technical problems related to the causes and effects of desertification.

Michael H. Glantz, scientist in the Environmental and Societal Impacts Group at the National Center for Atmospheric Research, has an extensive background in engineering, political science, and international relations. His most recent publication is *The Politics of Natural Disaster: The Case of the Sahelian Drought.* Dr. Glantz holds a Ph.D. from the University of Pennsylvania and has taught at the Universities of Pennsylvania and Colorado and at Swarthmore and Lafayette Colleges. He is a member of Colorado Governor Lamm's Drought Council.

Photograph by Carl Purcell, Washington, D.C.

Desertification

Environmental Degradation
in and around Arid Lands

edited by Michael H. Glantz

with a Foreword by Viktor A. Kovda

Westview Press
Boulder, Colorado

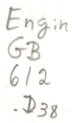
Westview Special Studies in
Natural Resources and Energy Management

Copyright © 1977 by Westview Press

Published in 1977 in the United States of America by
  Westview Press, Inc.
  1898 Flatiron Court
  Boulder, Colorado 80301
  Frederick A. Praeger, Publisher and Editorial Director

Library of Congress Cataloging in Publication Data
Main entry under title:
Desertification.
  Includes bibliographical references and index.
  1. Desertification.   I. Glantz, Michael H.
GB612.D38   910'.02'154   77-3901
ISBN 0-89158-115-4

Printed and bound in the United States of America

To my daughter, Mica, and her friend, S. B.,
and to my brother, Ronnie, and sisters,
Sally and Patsy

Contents

Foreword

On vast areas of the land droughts occur ever more often. Resources of the fresh surface and ground waters decrease noticeably. The erroneous methods of the utilization of natural resources by man—complete clearing of the forests and shrubs on the vast areas, overgrazing and destruction of the grass sod by animals, ploughing of the slopes, monocultural farming practices and regular burning of vegetation—have aggravated the general unfavorable tendencies of the land aridization in our time. The originally fixed sands of savannas, steppes, and semideserts begin to shift, and creep over the fields, roads, and wells. Dust storms destroy the soils of the cultivated fields and ruin the crops. Water erosion destroys the arable humus layer of the soil and its fertility. The albedo of the stripped soil increases multifold. The water content in the soil goes down, whereas the surface evaporation goes up.

The processes of the accumulation of toxic salts in the soils and natural waters gain speed. The vegetation cover deteriorates. The former forest landscapes change into savannas and steppes. The savannas and steppes look more and more like dry steppes whereas the dry savannas and steppes, and the meadows, turn into deserts, semideserts, and saline plains. The accelerating process of land desertification takes place in Africa and South America, in Asia and North America. The biological productivity of the soil goes down, the field crops and the herbage perish. Hundreds of thousands and even millions of people living in these areas suffer from malnutrition and hunger caused by the desertification processes and droughts. In the countries where people's interests are guaranteed by the government the burden of the droughts and desertification processes is not so great. Occasionally, however, socioeconomic conditions—the immense income gap and low standards of living of the poorer sections of the population neglected by the government—combine with drought and desertification to cause a social disaster. That is why the natural, economic, and social aspects of desertification became the focus of attention of progressive world public opinion and in 1977 were included in a representative intergovernmental conference of the United Nations member countries to discuss the problems of desertification and work out a plan of controlling this phenomenon and its aftereffects.

A profound analysis of the causal mechanism of the desertification processes is given in *Desertification*, edited by Michael Glantz, to which experts in various fields of science and representing the experience of different countries have contributed. Much attention is paid to the evaluation of the adverse changes in the climate and weather that result in desertification. The role of the oceans and of the man-induced ecological violations of nature are analyzed as well as the importance of local national policies and the technological level of agriculture. The authors of the book avail themselves of every opportunity to give valuable practical recommendations, to share the experience gained in controlling desertification and in reclaiming destroyed territories—the experience of Somalia, the U.S.S.R., and other countries. The book gives a detailed treatment of the hard lessons learned as desertification advanced and prolonged droughts occurred in the Sahel region of West Africa.

It highlights the role of the United Nations in the struggle against the desertification processes. It abounds in new facts and generalizations, and will be a good manual for many practical workers.

Viktor A. Kovda

Preface

Although attention has been drawn to desertification as a direct result of the recent extended drought in the West African Sahel, it is important to note that researchers throughout the world have been concerned with desertification and desertification-related problems at least since the beginning of the twentieth century. In many instances their findings as well as their warnings have gone unheeded by decision makers, political and otherwise. It is hoped that the problems related to desertification will be depoliticized, both nationally and internationally, and that in view of the United Nations Conference on Desertification, a utopian view of "harmony of interest" among states within the international community might prevail.

This book represents a collection of papers on the subject of the creation of desert-like conditions as a result of human or natural causes. The contributions are from disciplines in both the natural and the social sciences, and were designed to represent the varying levels of analysis—local, national, regional, and global—as well as geographic scope. Considerable reference material has been provided throughout the book for those interested in pursuing the various subtopics in greater depth.

I would like to acknowledge the encouragement and support of the following people: Dr. M. V. Sanin and Dr. Nikolai S. Orlovsky, Desert Research Institute, Turkmen S.S.R.; Mr. Mahalmadane Djiteye and Mr. Abdullah Sow, Government Research Station, Sotube, Mali; Professor A. T. Dick Grove, Downing College, Cambridge University; Dr. E. G. Davy, Consultant, World Meteorological Organization; Ms. Patricia Paylore, Arid Lands Study Center, University of Arizona, Tucson; Dr. Mohamed El Khawas, Federal City College, Washington, D.C.; Mr. Harry van Loon, National Center for Atmospheric Research, Boulder, Colorado.

I would like to thank Elmer Armstrong and Maria Krenz for their editorial assistance and Betty Wilson, Betsy Holdsworth, Ursula Rosner, Gloria Parker, and Carol Shea for their assistance.

I would like also to acknowledge gratefully the support of the

International Federation of Institutes for Advanced Study (Stockholm), the Rockefeller Foundation (New York), the Aspen Institute for Humanistic Studies (Colorado), and the National Center for Atmospheric Research (NCAR). NCAR is operated by the nonprofit University Corporation for Atmospheric Research under sponsorship of the National Science Foundation.

The views expressed in these chapters are those of the individual contributors and do not necessarily reflect the views of the supporting organizations.

Michael H. Glantz

The Contributors

A. G. Babayev is president of the Turkmenistan Academy of Sciences. He is one of the founders of the Desert Institute, Ashkhabad, U.S.S.R.

Erik P. Eckholm is a senior researcher with Worldwatch Institute, a private, nonprofit organization in Washington, D.C., devoted to the analysis of emerging global issues. A graduate of the Johns Hopkins School of Advanced International Studies, Dr. Eckholm is author of *Losing Ground: Environmental Stress and World Food Prospects* and coauthor of *By Bread Alone*. He is now writing a book on the impact of the environment on human health, the research for which is supported by the United Nations Environment Program.

Michael H. Glantz, scientist in the Environmental and Societal Impacts Group at the National Center for Atmospheric Research, has an extensive background in engineering, political science, and international relations. His most recent publication is *The Politics of Natural Disaster: The Case of the Sahelian Drought*. Dr. Glantz holds a Ph.D. from the University of Pennsylvania and has taught at the Universities of Pennsylvania and Colorado and at Swarthmore and Lafayette Colleges. He is a member of Colorado Governor Lamm's Drought Council.

H. N. Le Houérou is the director for environmental sciences at the International Livestock Center for Africa in Addis Ababa, Ethiopia, and is also the scientific coordination officer for UNESCO's Man and the Biosphere Project Number 3, which is concerned with the impact of human activities on grazing lands. Until August 1975 he served as the senior officer-in-charge for the Grasslands and Pastures Group in the Plant Production and Protection Division at the United Nations Food and Agriculture

Organization in Rome. He has written extensively on desertization and desert reclamation and is the author of *Desert Ecology and Desertization*.

Richard W. Katz, a scientist with the Environmental and Societal Impacts Group at the National Center for Atmospheric Research (NCAR), received a Ph.D. in statistics from the Pennsylvania State University. He has worked at the Center for Climatic and Environmental Assessment of the National Oceanic and Atmospheric Administration in Columbia, Missouri, and as a postdoctoral fellow at NCAR. Dr. Katz's research includes work with probabilistic models for precipitation and statistical crop-weather models.

William W. Kellogg is a senior scientist in the Climate Project at the National Center for Atmospheric Research. He has been president of the American Meteorological Society (AMS) and was a member of the AMS delegation to the People's Republic of China in 1974. Dr. Kellogg is currently a member of the National Academy of Sciences Committee on Scholarly Communication with the P.R.C.'s Panel of Atmospheric and Earth Sciences.

Viktor A. Kovda is director of the Institute of Agrochemistry and Soil Science at the U.S.S.R. Academy of Sciences and is professor of soil science at Moscow State University. He has served as a member of the executive council of the International Council of Scientific Unions (ICSU) and has been the president of its Scientific Committee on Problems of the Environment (SCOPE). Dr. Kovda is author of several books and articles in the field of soil science.

Jeremy Swift is a research fellow in the School of African and Asian Studies at the University of Sussex (Brighton, England). He has worked for the United Nations Food and Agriculture Organization and for the International Union for the Conservation of Nature and Natural Resources, studying problems of wildlife and natural resource conservation. He is currently researching the ecology and economics of nomadic pastoralism, with emphasis on development strategies.

J. Dana Thompson is a staff scientist at JAYCOR, a private research firm in Alexandria, Virginia. After receiving a Ph.D. in meteorology from Florida State University, he was a postdoctoral fellow at the National Center for Atmospheric Research. At present, Dr. Thompson is engaged in studies of the dynamics of the Somali Current, currents in the equatorial and eastern Pacific, and the environmental impact of ocean thermal power plants.

James T. Thomson is an assistant professor in the Department of Government and Law at Lafayette College, where he has been teaching since 1974. Dr. Thomson has spent four years in Niger, including two years of field research on traditional court systems, dispute resolution, and development in a rural area of south-central Niger.

Stephen H. Schneider, deputy head of the Climate Project at the National Center for Atmospheric Research since 1973, holds a Ph.D. in plasma physics from Columbia University. Dr. Schneider is editor of the journal *Climatic Change*. His most recent publication is *The Genesis Strategy* (with Lynn Mesirow).

Helen Ware is a research fellow in the Department of Demography of the Australian National University. As field director of the Changing African Family Project, a collaborative enterprise incorporating research institutions in twelve African countries, she has traveled and worked extensively in Africa. Dr. Ware's paper on lessons to be learned from recent drought experience was published by the Special Sahelian Office of the United Nations.

1

The U.N. and Desertification: Dealing with a Global Problem

Michael H. Glantz

The United Nations Conference on Desertification will have been at least the seventh major international conference since 1970 designed to deal with a "global" problem. Judgments about the relative success or failure of such gatherings in meeting their implicit and explicit goals have varied widely. Often such evaluations have been affected by the existence of cleavages within the international community; cleavages based on political, cultural, ideological, and economic factors. Given the existence of these factors, there are often differing views on what such international conferences might achieve, on who should bear the financial burden of carrying out the suggestions of the conference, and on the nature as well as the causes of the problems being evaluated (de Gaspar, 1976).

While one can make a list of functions that these conferences might fulfill,[1] it can also be shown that for many of the conference participants and observers there will still exist a crisis of expectations with respect to the outcomes of the conference. It is the purpose of this chapter to make explicit some of the potential political and cultural obstacles to international programs designed to combat desertification. It is hoped that making these obstacles explicit will be of use to political and other decision-makers in shaping their expectations about the results of the conference. This will perhaps help them to distinguish between what is desirable and what is possible, and to distinguish between what is desirable and what is essential (Morgenthau, 1969).

Definitions of Desertification

The development of desert-like conditions where none had existed before has generated two different terms to define it—*desertization* and *desertification*. The distinction between these terms is primarily based on the location of the extension of desert-like conditions. For example, Le Houérou, among others, has defined desertization as "the extension of typical desert landscapes and landform (i.e., those regions receiving between 50-100 and 200-300 mm)" (see chapter 2). Le Houérou intentionally excluded from this definition the same process as it occurs in regions of higher rainfall which are not adjacent to desert areas.

Citing the geographic limitations of Le Houérou's definition, Rapp (1974) expanded it to include higher rainfall areas. He defined desertization as "the spread of desert-like conditions in arid or semiarid areas up to 600 mm, due to man's influence or to climatic change." Rapp also noted that his definition conformed with that of desertification, as defined by Sherbrooke and Paylore (1973). Thus, the two terms have been used by different authors sometimes to describe different processes and sometimes the same process. To the United Nations, the problem of definition was resolved when the United Nations Environment Program (UNEP), the agency within the United Nations system with responsibility for the Conference on Desertification, chose desertification to apply to the process as defined by Rapp.[2] (In this book Rapp's definition will be used to describe desertification unless otherwise specified by the individual contributors.)

In addition to the problem of definition, it is important to note that there also exists some controversy concerning the causes of desertification, that is, whether they are to be attributed to anthropogenic factors or to natural (physical) ones. Many authors seem to agree with Novikoff's view (1975) that "the main agent for desertisation is actually man."

Cycles of Interest in Desertification

It is well accepted that recent interest in the process of desertification has been a result of the social, political, and environmental implications of the recurring droughts and dry spells which occurred in the West African Sahel between 1968 and 1973. It has been estimated that during this period the drought directly affected 6 million of the region's inhabitants and 25 million cattle, leading to an estimated 100,000 human deaths and up to 40% loss of cattle. The cattle statistics, however, remain quite controversial.

In addition, before the cattle perished, large areas (especially around watering points) were denuded of vegetation as a combined result of the following factors: excessively large number of cattle on a dwindling vegetation resource; breakdown of traditional grazing patterns due to the construction of deep wells; pressures created by cultivators seeking more land to farm in the north, especially during the preceding 15-year period of above average rainfall conditions. For at least four of the six Sahelian states—Niger, Chad, Mali, and Upper Volta—the rangelands constitute one of the few potential economic resources that, given proper management, might be developed.

The recent Sahelian drought was the third major one of this century. The literature shows that scholars and governments have been dealing with drought-related problems, such as desertification, as well as with drought-related social problems throughout the century. Yet the acuteness of the problem was highlighted recently by the accelerated loss of potentially productive soil to desert-like conditions, conditions which, according to Cloudsley-Thompson (1971), are "primarily due to the deliberate action of man, in many instances within living memory." The loss of this land came at a time when world food productivity was declining for the first time in 20 years and populations were still increasing (Brown, 1974). In addition, 1972 was a year of anomalous weather, including a major drought in the Soviet Union. The following map (figure 1.1) shows the geographic scope of the anomalies. The drought caused the Soviet Union to make unusually large grain purchases from American grain corporations, leaving less available for food relief in other areas (Trager, 1975). The energy crisis, too, had an adverse impact on the ability of the Sahelian states to receive food relief during the drought situation. All of these factors represent major obstacles to regional economic and political development and, in some instances, actually threaten the economic viability of several countries. The gravity of the situation was brought out when it was recently suggested that "if desert spreading is not stopped, two African countries will completely disappear within the next decade" (U.N., 1976a).

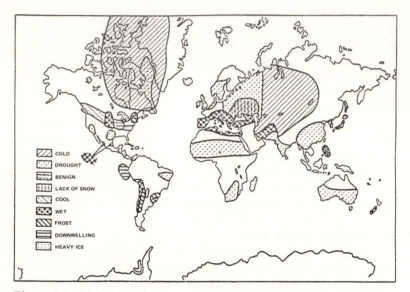

Figure 1.1. Global weather anomalies 1972. From McKay and Allsopp, 1976.

Definition of a Global Problem

The occurrence of the recent Sahelian drought, with all its social and political ramifications, has served as the catalyst to elevate desertification to the status of a "global" problem. Other problems on the U.N.'s list of officially designated global problems are: the environment (Stockholm, 1972); population (Bucharest, 1974); food (Rome, 1974); women (Mexico City, 1975); human settlements (Vancouver, 1976); water resources (Buenos Aires, 1977); and technology transfer (1979). These problems are not necessarily perceived to be of equal importance by the various members of the international community and, while officially designated as global, are not necessarily treated with the same sense of urgency, commitment and/or interest. Russell and Landsberg (1971) seeking to categorize global problems defined them as "those problems that physically involve all or nearly all nations of the world, either as contributing parties (emitters) or as damaged parties (receptors) or both." Thus a problem can be global either in cause or in effect or in both. They also suggested other factors, which would be useful in categorizing global problems, such as time lag between cause and effect, the degree of uncertainty of the effects, the degree of reversibility of effects, types of problem linkages (social or physical), and the type of impact (pecuniary or nonpecuniary). Another author suggested that, in fact, there were very few truly global problems: "Environmental problems of a truly global nature affecting the biosphere as a whole are rather limited in number. They involve primarily the ocean and atmosphere—the 'commons' which are shared by all are the basis of our life-support system" (Quigg, 1973). How then does the problem of desertification fit into these taxonomies?

Global Problems and Desertification

As suggested in the preceding sections, problems can be considered global for a variety of reasons, with each reason commanding varying degrees of interest, urgency, and support from members of the international community. Desertification, however, does not fit neatly into either of the broad categories as mentioned by Quigg or Russell and Landsberg, and, in fact, it possesses its own combination of characteristics.

For example, desertification as an environmental problem is *primarily* national in cause and national in effect. One might easily argue that despite such secondary effects as Saharan dust blowing across the Atlantic and despite the migrations of refugees across international borders, desertification is a national problem in which the inhabitants of a country exacerbate the preconditions for desertification through such activities as land mismanagement, overgrazing, deep-well drilling, deforestation, and firewood gathering. Reinforcing this view are the growing number of satellite photos depicting different patterns of land use including the adverse effects of land mismanagement practices up to but not across international frontiers (Otterman, 1974, 1977; Rapp, 1974, 1975). Figures 1.2 and 1.3 show such situations.

One might effectively argue that unlike such problems as environmental pollution, often referred to as the "pollution of affluence" (Kay and Skolnikoff, 1972; Enloe, 1975), another characteristic of desertification as a process is that it can be viewed as classless. Although attention has focussed on its devastating impact in the poorer countries—such as those in sub-Saharan Africa, the Middle East, Southwest Asia, the Indian subcontinent, and in South America—the more developed and the richer nations

Figure 1.2. Satellite image of boundary zone in South Africa (left) and Lesotho (right). The three main zones are, from left to right: (1) predominantly grey pattern—wheat fields and other vegetation in South Africa; (2) pale zone—overgrazed, eroded and vegetation-poor farmlands in Lesotho, separated from South Africa by a boundary fence; (3) dark zones to the right—mountains and plateaus in Lesotho with green mountain pastures. Date of ERTS image is November 21, 1972. Reproduction by W. Arnberg, Stockholm. This clearly shows the countrywide extent of soil erosion in Lesotho as contrasted to the high-economy conservation farming in South Africa. Note the pale zone of soil degradation in the Orange River Valley to the right (Rapp, 1975).

have also been adversely affected by the problem. However, the more affluent and the more developed states have been able to divert at least some of their resources into programs designed to combat desertification. Programs have been designed to stop the loss of topsoil or to reclaim lands already lost by resorting to weather and climate modification schemes. The poorer states, on the other hand, do not have the resources available for the develop-

Figure 1.3. A plainly visible division between grazed and ungrazed lands in the Negev area of the Sinai desert. The photo, according to Otterman, shows the denuded, high albedo (reflectivity) regions of the Sinai and Gaza Strip in sharp contrast to the darker western Negev. The division coincides with the 1948-49 Armistice Line between Israel and Egypt, where a fence was built in the early 1970s (NASA photo; Otterman, 1974, 1977).

ment and the implementation of such programs, and in many cases also lack the technical expertise required to develop, enact, and oversee such programs. In addition there exists the problem of political instability within these states, which forces leaders to divert much time, energy, and resources to matters relating to the maintenance of security. Such instability also disrupts continuity with respect to economic and political development plans, under which the problem of desertification falls. The administrative problem associated with managing the rangelands in sub-Saharan Africa during times of relative peace and stability have been well exposed by Baker in his paper, "The Administrative Trap" (1975).

As one can see, though the problem of desertification might be considered classless, the ability to cope with it cannot be considered classless.

The one aspect of desertification that is common to the other officially designated global problems is the widespread interest that exists among the individual members of the international community. It is this collection of individual national interests that makes desertification a global problem. Part of this interest results from the fact that one-ninth of the surface of the earth is affected by desertification, that up to one-third of the surface of the earth has already been classified as arid or semiarid, and that at least half the countries in the world have been affected to some degree by the problems associated with aridity (Development Forum, 1976). Considering these factors together, desertification has become at least temporarily "one of the most serious environmental problems" confronting the international community.

Yet, despite the generally recognized importance of the problem, levels of commitment within the international community are sure to vary, partly due to how the states perceive desertification as a problem, partly due to the perceived lead time needed to deal with the problem, and partly due to the eventual passing of the crises linked to the recent Sahelian droughts. Once the crisis is perceived to have passed, interest in resolving the perennial problems linked to desertification dissipates, even though these underlying causes of desertification still remain. Also, UNEP recently noted that an awareness of desertification will not necessarily mean that effective measures to deal with it will eventually be developed and implemented. This is so, given that there exist both within the national and international communities potential political and cultural obstacles to the implementation of programs which might effectively combat the desertification process. Along the same lines one reporter, commenting on the World Food Conference, observed that "as often happens with conferences, the crisis atmosphere that generated them wears off and good purposes become bogged down in institutional bureaucracies and political expediencies" (*Christian Science Monitor*, 1975). The following section will briefly refer to some of the potential sociopolitical obstacles.

The International Setting

During discussions within the United Nations on desertification, differing views were expressed on the need and on the desirability of a world conference on the subject. On the one hand, representatives of developing states, while recognizing that the problems differ from one country to the next, maintain that desertification could not be solved nationally or regionally but must be attacked through a global strategy. It has also been argued that a global strategy would be most appropriate given that those states most affected by the problem are also those states which lack the financial as well as the technical resources needed for effectively combating desertification. On the other hand, representatives from some of the more developed states, while agreeing on the need for "an integrated world programme in the field of research and development and application of science and technology to the scourge of desertification" (United Nations General Assembly [U.N.G.A.], 1974), disagreed on the need for or effectiveness of a global conference. For example, Neufeldt of the Federal Republic of Germany commented that "the mere creation of a new institution would not, however, suffice to solve the problem; what was also required was the political will to take effective action at the national and regional levels." A French representative (U.N.G.A., 1974) noted that "the U.N. system was already fully capable of considering all aspects of the problem of desertification" and that the proposed conference would not make a significant contribution.

This divergence of opinion reflects the desire of the developing states to seek international recognition of some of their problems in order to achieve the international cooperation needed to resolve them. At the same time it reflects on the part of the developed states a preference for more concrete measures to be taken at the national and regional levels. Given these differences along with others that exist in the international community it seems to be a difficult task for any group to create a worldwide strategy to combat either desertification or any of the other officially designated global problems. It appears from an international perspective that desertification is destined to remain an insurmountable problem, unless the international community can foster a harmony of interest at least long enough to deal with this particular global problem.

In the late 1930s E. H. Carr, discussing international politics in the interwar years, focussed on the differing perceptions that existed with respect to the concept of the harmony of interest. To the "realists," the harmony of interest is based on "what is good for my country will be good for the rest of the international community." This view places the national interest above the interests of the community. On the other hand, the "idealists" say "what is good for the international community will be good for my country." Seen from the perspective of the realists in the non-affected areas, desertification as a global problem does not exist and therefore the members of the international community should assist in programs to combat it only if they deem it to be in their national interest. To the idealists, however, desertification is a problem that merits concern and involvement from the global community, based on the belief that what is good for the international community will be good in the long run for the various nations which make up that community. Thus, it is an idealist's view of the harmony of interest that is needed to prevail for the global community to develop and to implement effectively programs designed to combat desertification. Unfortunately, as one author noted, "the solutions to the problem of desertification often depend on political decisions" (U.N., 1976b). Such a view applies not only to the international setting but to the national setting as well.

Finally, it should be mentioned that there are also conflicting views on the level of interdependence that exists within the international community. Lester Brown, for example, has suggested that there is a high level of interdependence. In *The Interdependence of Nations*, Brown (1972) expressed the following view: "When one inventories the many kinds of ties now existing among nations, one begins to appreciate how rapidly our daily well-being is becoming irrevocably dependent on the resources and cooperating of other nations." Yet such a view overlooks the fact that there is also a high degree of noninterdependence among states, and hence it can lead to misperceptions about what states are likely to do as opposed to what they ought to do with respect to assisting other nations (Glantz, 1977).

Other writers (Waltz, 1970; Volkov, 1976) have noted the myths of interdependence. While Waltz stressed the view that the major actor in the international system is still the state, Volkov took more of an ideological stance, stressing that "governments of industrial capitalist countries have of late adopted the conception of interdependence as an official theory. Spokesmen of these countries cite interdependence to justify their government's egotistical policy toward the developing countries at international conferences and in the United Nations." Awareness of these conflicting views, also referred to by Carr (1964), can be useful in understanding international responses to the problems related to desertification.

The National Setting

As an environmental problem which is primarily national in cause and national in effect, it can be argued that solutions to the problems of desertification will ultimately rest with national decision-makers and will directly and primarily depend on national political, economic, and social conditions within those states directly affected by desertification. These sociopolitical factors are reinforced by other constraints of a more technical nature (Le Houérou, 1975; UNEP, 1975a). The crises in the West African Sahel can be considered as an example.

The Sahelian states are confronted by a hydra-headed internal crisis made up of elements such as those suggested by Hilling (1973): geographical, technical, and institutional constraints; by Pye (1966): participation, identity, legitimacy, distribution, integration, penetration crises; by Lofchie (1971): embryonic political institutions carrying the burdens of welfare and development and the implications of a rapid "transition from labor intensive to automated industrial systems"; by Wallerstein (1971): "the weakness and fragility of the state machinery" and "the perennial shortage of state revenue"; by Hardin (1972): the tragedy of the commons; and by Baker (1975): bureaucratic jurisdictional rivalries and disputes.

The Sahelian states are among the poorest in the world. Often overwhelmed by these crises, they are, therefore, in need of financial, technical, moral, and other support, if they are to have any chance whatsoever to cope effectively with the problems that are linked to desertification within their borders. One must not forget that recent attention to desertification-related problems was the result of their plight during the extended drought situation between 1968 and 1973, and that with the return to favorable rainfall, international—and in some cases national—interest in dealing with desertification will generally disappear. As Baker (1976) noted, "there is . . . a great danger that when the rains come in the Sahel, and the millet grows again, then 'the problem' will be considered over until next time."

Of all the crises confronting developing countries like those of the Sahel, it can be argued that political instability deserves the most attention. Supporting this view, Lofchie wrote that "the principal justification for emphasizing political rather than other constraints is the extent to which political will and political action can stimulate development in spite of other countervailing obstacles" (1971). Le Houérou (1975), too, suggested that "technically desertization can be halted, there is no question about it, and many desertized areas could be restored to production. However, the medicine can only be applied if there is a strong political will and power to do so (which in turn supposes political stability) since the actions to be taken closely interfere with the daily life of the population concerned." Perhaps it is the existence of these political constraints that explains, at least in part, why much of the scientific information which has been developed by researchers in offices, in laboratories, and in the field, is either not being used at all or is not being used for the benefit of society. Such existing information includes an awareness of the detrimental effects on the environment and eventually on the economy of deforestation, indiscriminate deep-well construction, land mismanagement practices, livestock rationalization, and similar practices.

Experts have noted and discussed at great length these questions as well as others related to desertification. They also proposed approaches to the resolution of some aspects of these problems long before the occurrence of the recent Sahelian drought. However, little has been done throughout the twentieth century in a concerted way to rid this region or others like it of its chronic problems. Therefore one should not take for granted that the convening of a world conference will by itself contribute significantly to the way in which chronic problems, such as desertification, will be resolved. It can be argued that some solutions have been recognized for a long time but that nothing has been done to implement them (UNEP, 1975; Glantz, 1976).

In summary, expectations about how this world conference confronts desertification should be based on what the political reality is and not on what it ought to be (Carr, 1964). But, perhaps, in the long run such conferences, aside from fulfilling such functions as the sharing of information, views, and expertise on a particular problem, will move states in the international system at least one step closer to Carr's idealist's conception of the harmony of interest.

Notes

1. According to the U.N. General Assembly, the Conference on Desertification was designed to draw the attention of the international community to the magnitude of the problem of desertification on all continents and to mobilize all available knowledge and experience in preparing a program of action to combat it (U.N. General Assembly Records, 1974).

2. "It was the understanding of the UNEP Executive Director [Mostafa K. Tolba] that the scientific community frequently uses 'desertification' to mean a process caused by man, and 'desertization' a process regardless of cause. For the purposes of carrying out this [U.N.] resolution, therefore, the Executive Director would interpret 'desertification' as having the broader meaning often attributed to 'desertization' " (UNEP, 1975b).

References

Baker, R., 1975. The Administrative Trap: Conflict Between Administration and Ecology. University of East Anglia School of Development Studies, discussion paper no. 5.

Baker, R., 1976. Innovation, technology transfer and nomadic pastoral societies. In *The Politics of Natural Disaster: The Case of the Sahel Drought*. M. H. Glantz, ed. New York: Praeger Publishers. Pp. 176-185.

Brown, L., 1972. *The Interdependence of Nations*. New York: Foreign Policy Association Headline Series no. 212, October.

Brown, L., and E. P. Eckholm, 1974. *By Bread Alone*. New York: Praeger Publishers.

Carr, E. H., 1964. *Twenty Years Crisis*. New York: Harper & Row. Originally published in 1939.

Christian Science Monitor, 1975. Food: a year after Rome. 28 November 1953.

Cloudsley-Thompson, J. L., 1971. Recent expansion of the Sahara. *International Journal of Environmental Studies* 2 (38).

de Gaspar, Diogo, 1976. Beyond conference ritual. *Development Forum*, U.N., 4 (6) July-August, pp. 1-2.

Development Forum, U.N., 1976. United Nations conference on desertification in 1977. 4 (6) July-August, p. 6.

Enloe, C., 1975. *The Politics of Pollution in a Comparative Perspective*. New York: David McKay Co.

Glantz, M. H., 1976. Nine fallacies of a natural disaster. In *The Politics of Natural Disaster: The Case of the Sahel Drought*. M. H. Glantz, ed. New York: Praeger Publishers. Pp. 3-24.

Glantz, M. H., 1977. The value of a long range weather forecast for the West African Sahel. *Bulletin of the American Meteorological Society*, 58 (2), pp. 150-158.

Hardin, G., 1972. Tragedy of the commons. *Exploring New Ethics for Survival*. Baltimore, Md.: Penguin Books.

Hilling, D., 1973. West Africa's land-locked states: some problems of transport and access. In *Drought in Africa*. D. Dalby and R. J. Harrison Church, eds. London: University of London School of Oriental and African Studies.

Kay, D., and E. B. Skolnikoff, eds., 1972. *World Eco-crises: International Organizations in Response*. Madison: University of Wisconsin Press. P. 310.

Le Houérou, H., 1975. Science power and desertization. In *Proceedings of the I.G.U. Meeting on Desertification*. Cambridge: Cambridge University Department of Geography. 22-26 September, p. 26.

Lofchie, M., ed., 1971. *The State of Nations: Constraints on Development in Independent Africa*. Berkeley: University of California Press. Pp. 5-6.

McKay, G., and T. Allsopp, 1976. Global interdependence of the climate of 1972. In *Proceedings of the Mexican Geophysical Union Symposium on Living with Climate Change*. Mexico City, May.

Morgenthau, H. J., 1969. *A New Foreign Policy for the United States*. New York: Praeger Publishers. P. 10.

Novikoff, G., 1975. The desertisation of range lands and cereal cultivated lands in Presaharan Tunisia. In *Tunisian Presaharan Project Progress Report #3*. G. Novikoff, F. H. Wagner, and Prof. Skouri, eds. Logan, Utah: Utah State University. February, p. 163.

Otterman, J., 1974. Baring high-albedo soils by overgrazing: a hypothesized desertification mechanism. *Science* 96:531-533.

Otterman, J., 1977. Anthropogenic impact on the albedo of the earth. *Climatic Change* 1 (2).

Pye, L., 1966. *Aspects of Political Development*. Boston: Little, Brown and Co. Pp. 62-67.

Quigg, P. W., 1973. *Environment: The Global Issues*. New York: Foreign Policy Association Headline Series no. 217. October, p. 4.

Rapp, A., 1974. A Review of Desertization in Africa: Water, Vegetation and Man. Stockholm: Secretariat for International Ecology, Sweden (SIES).

Rapp, A., 1975. Soil erosion and sedimentation in Tanzania and Lesotho. *Ambio* (Stockholm) 4 (4), pp. 154-163.

Russell, C. S., and H. H. Landsberg, 1971. International environmental problems: a taxonomy. *Science* 172: 1308.

Sherbrooke, W. C., and P. Paylore, 1973. *World Desertification: Cause and Effect*. Tucson: University of Arizona Office of Arid Lands Studies: Information Paper no. 3, pp. 1-5.

Trager, J., 1975. *The Great Grain Robbery*. New York: Ballantine Books.

U.N., 1976a. United Nations Conference on Desertification. OPI/NGO/SB/96. 20 May, p. 1.

U.N., 1976b. Press release, HE/340. 8 April.

UNEP, 1975a. *Overview in the Priority Subject Area: Land, Water and Desertification*. Nairobi: UNEP/Prog./2, February.

UNEP, 1975b. International cooperation to combat desertification. Geneva: UNEP Liaison Office, *Ad Hoc Interagency Task Force*, 26-28 June, p. 2.

U.N.G.A., 1974. U.N. General Assembly (XXIX). Second committee, 1636th meeting. 22 November, p. 369.

Volkov, M., 1976. *The Strategy of Neocolonialism Today*. Moscow: Novosti Press Agency Publishing House. Pp. 30-46.

Wallerstein, I., 1971. The range of choice: constraints on the policies of governments of contemporary African independent states. In *The State of Nations*. M. Lofchie, ed. Berkeley: University of California Press. Pp. 5, 19-33.

Waltz, K., 1970. The myth of national interdependence. In *The International Corporation*. C. Kindleberger, ed. Cambridge, Mass.: M.I.T. Press. Pp. 205-233.

2

The Nature and Causes
of Desertization

H. N. Le Houérou

The word *desertization* is preferred to the term *desertification* for several reasons. Desertization has been defined (Le Houérou, 1962, 1968) as the extension of typical desert landscapes and landforms to areas where they did not occur in the recent past. This process takes place in arid zones bordering the deserts under average annual rainfalls of 100 to 200 with outside limits of 50 to 300 mm. The word desertification has been—and still is—used by many authors to describe degradation of various types and forms of vegetation, including the subhumid and humid forest areas, which have nothing to do with deserts either physically or biologically.

The word desertization is being used by an increasing number of modern ecologists and other scientists (Rapp, 1974; Floret and Le Floch, 1973; Ionesco, 1972; Monod, 1973; Wagner, 1974; Novikoff, 1974; Baumer, 1973, 1974) and also by international organizations: Food and Agriculture Organization (FAO), United Nations Educational, Scientific, and Cultural Organization (UNESCO), UNEP, International Livestock Center for Africa (ILCA). For example, a conference was held in 1975 in Teheran by UNEP and the Iranian Government on "de-desertization" and a world conference on the subject (U.N. Conference on International Coopera-

This paper appeared in the *Proceedings* of the International Geographical Union meeting on Desertification, Cambridge University, England, 23 September 1975.

tion to Combat Desertification) will be held in 1977 under the aegis of the U.N. and UNEP.

The Desert and Its Limits

The *desert*, as it is understood here, is characterized by eremitic vegetation where perennials are distributed on a contracted pattern (Monod, 1954) on the regs[1] (gravelly or pebbly plains) and hammadas (flat, stony desert pavements of flagstones on a horizontal geological structure). However, desert dunes, sandy areas, wadi terraces, and water spreading zones (maaders) may have a diffuse vegetation pattern, even in the true desert, such as fezzan. This concentration of perennial vegetation along the stream network occurs along the 100-mm isohyet (line of equal precipitation) both north and south of the Sahara as well as in the Near and Middle Eastern deserts.

Many more complex definitions of deserts based upon bioclimatic empiric formulas have been proposed. When put on a map, most of the index values given as desert limits tend to run along the 100-mm isohyet (Le Houérou, 1959, 1968, 1969). There are some exceptions, however, especially in the coastal deserts of Chile and Peru and in Namibia (currently referred to as South West Africa) where, owing to high air moisture during part of the year, the upper limit of the desert seems to be approximately the 50-mm isohyet. These are also referred to as "mist deserts."

As a matter of fact, this limit of 100 mm corresponds to a permanent water deficit, a situation in which the potential evapotranspiration is higher on the average than precipitation each month of the year. Annual potential evapotranspiration is on the order of magnitude of 500 mm or more, or at least 15 times higher than rainfall. The *arid* zone is understood here to be those areas lying between the 100- and 400-mm isohyets north and south of the Sahara and in the Near and Middle East. This holds also for the arid zones of North and South America (still with the exception of coastal arid zones such as northern Chile and Namibia where the arid zone lies between the isohyets of 50 and 300 mm). The arid zone is characterized by vegetation distributed on a diffuse pattern and usually referred to as steppic. This is, for instance, the case of the steppes of North Africa, the Near and Middle East, and also of the Sahelian zone south of the Sahara as well as in East and South Africa. South and north of the Sahara and in the Near and Middle

East, the 400-mm isohyet corresponds roughly to the lower limit of productive and regular farming of food and industrial crops without irrigation (wheat and barley in North Africa and the Near East; millet, sorghum, cowpea, and ground nuts in the tropics). The 100- to 400-mm belt, or arid zone, or steppic zone, is basically a pastoral area with some exceptions in Tunisia and Libya, where fruit trees are grown with dry farming techniques. Owing to heavy demographic pressures, however, there is now a progressive encroachment of cropping on the pastoral areas which, as we will see, is one of the major causes of desertization.

It is therefore understood that many so-called "deserts" are not true deserts but are simply arid zones; for example, central Australia, Kalahari, northern Mexico, and the southwestern United States. However, the Sahara, Sinai, Arabian Peninsula, Iranian, and some central Asian deserts are true deserts along with Atacama, west Namibia, and perhaps some limited areas in North America (Death Valley). One can hardly call "desert" the beautiful olive groves of southern Tunisia and Tripolitania developed with dry farming techniques under only 150 to 200 mm of rainfall, which is similar to the rainfall occurring in the most arid parts of Australia, Kalahari, and Sonora.

The Nature of the Phenomenon

The creation of new desert landforms and landscapes such as regs, hammadas, barkhans, nebkas, reboubs, and other types of continental dunes is a result of wind erosion. On the margin of deserts, these newly established superficial land forms are superimposed on an inherited typical arid zone morphology carved by water erosion and sedimentation and characterized, among other features, by interlocked pediments in the synclines. The denudation of pediments of all perennial vegetation (except along the drainage network) gives way to intense wind erosion which, in a very few years, results in the formation of typical regs or hammadas; the coarser materials (0.05 to 2.0 mm in diameter) are blown away and accumulate in barkhans, nebkas, reboubs, sand veils, and other accumulation forms; the finer products are taken away in the wind, and move over continental distances to accumulate in the form of loess. Dust from the Sahara is known to cross the Atlantic Ocean and "sand-rains" are reported from as far afield as northern Europe (Le Houérou, 1973; Rapp, 1974; MacLeod, 1976). Two main types of the reg-making process are described by Monod (1937), autochthonous and allochthonous. Autochthonous regs are made *in situ* by disaggregation of geological outcrops (often limestones or Quaternary limecrusts). Allochthonous regs develop on alluvial material of terraces or colluvia where the finer detrital elements are blown away, leaving gravel and pebbles which cover the surface of the reg, protecting it from further wind erosion.

Where vegetation in an arid zone is not degraded, there is always a diffuse ground cover of at least 20 to 40% of perennial species such as shrubs, undershrubs, and perennial grasses. This is enough to protect the soil surface from erosion; or, more precisely, wind erosion in these conditions is compensated by sand deposited behind the obstacles that perennial plants constitute.

Where the distance between two perennial species is equal to or longer than five times their height, deflation is no longer compensated by sand deposition and erosion increases up to the point where the whole land surface is covered by pebbles or stones after removal of all movable material. The sand that drifted away accumulates in barkhans, reboubs, nebkas, sand veils, and the like. This sand drift takes place over rather short distances, seldom surpassing a few hundred meters and then up to no more than a few kilometers. The final result is that soil layers, which used to be superposed as sandy or loamy deposits over rock outcrops or limecrusts, become juxtaposed as a result of wind erosion. Permanent plant life then becomes impossible, for the lack of water reserves in the shallow soil of the reg prevents seedlings from surviving the first prolonged drought period. In shifting sand, perennial species, too, cannot get established. Just as soon as the seedlings have emerged, they are uprooted and blown away.

Another source of desertization is caused by the action of raindrops. Barren silty, loamy, or clay ground becomes sealed on the surface by the action of raindrops. Raindrops disperse the fine elements of surface soil aggregates. The particulates, thus freed, tend to fill up the pores in the soil surface. This surface becomes sealed (impermeable), and water intake becomes negligible with a runoff factor of 80% or more, depending on slope and degree of sealing. This leads to much drier soil conditions, resulting in the death of the remaining plant individuals, as was the case over numerous large areas of the Sahel during the 1968-73 drought.

A third mechanism of desertization is due to salinization and alkalinization, which are usually the results of faulty technology in water development schemes: using too salty waters on too heavy soils, insufficient drainage or even no drainage at all, using too much or too little water. This mechanism has led to the sterilization of huge areas—the so-called "scalds"—in North Africa, Egypt, Iraq, Iran, Pakistan, India, Australia, and other arid countries, contributing to the extension of man-made deserts.

The nature of desertization thus appears to be complex, with several different mechanisms ending in similar convergent results. It is always a physical, or chemicophysical, process, induced by biological, or biophysical, mechanisms which reduce plant cover and primary productivity. Eventually the environment becomes unusable for man and his livestock, thus narrowing the limits of the ecumene.

Is Desertization Irreversible?

It is not possible to answer this question with either a simple "yes" or "no." There is evidence that in some cases, especially on shallow soils with rainfall less than 200 mm, desertization is irreversible. For instance, in southern Tunisia the tracks left by the French, British, and German tanks during the battle of Ksar-Rhilane (February to March 1943) are still conspicuous in the landscape 30 years later (Le Houérou, 1959, 1969, 1974; Floret and Le Floch, 1973). The perennial, shrubby vegetation which covers this area (*Anthyllis henoniana—Gymnocarpos decander* association; Le Houérou, 1959) has not reestablished itself in spite of several exceptionally rainy years in the late 1950s. The average rainfall in this area is 80 to 100 mm and the substratum is soft senonian limestone overlaid by a thick limecrust, coated by a very thin and mobile sand veil. Similar facts have been observed in eastern Libya (Tobruk, Bir Hakeim), in western Libya (Bi Ayyad), and in many other places.

Wherever the ground is soft, light, and deep enough, vegetation recovery is possible, even in areas receiving as little as 60 to 80 mm of average annual rainfall; for example, in Mauritania (Monod, 1951; Adam, 1967) and in Libya, Tunisia, Iran (Le Houérou, 1959, 1962, 1965, 1969, 1973, 1975). In depressions benefiting from runoff, in water spreading areas, and in those areas having a water table, recovery may be excellent; but this is another story.

In what conditions degradation is or is not irreversible is far from clear. That is why in the definition of desertization I have used the locution "more or less irreversible," which to my mind means either irreversibility or having a very slow recovery potential, under current natural conditions. The extensive work of desert rehabilitation accomplished over several hundred thousand hectares in Iran (Shaidaee, 1974; Niknam and Ahranjani, 1975; Firouz, 1974, 1975) shows that wherever the soil is soft and deep enough, rehabilitation potentials (either naturally by protection from grazing and cultivation or artificially through shrubs planting or seeding) are quite good at a relatively low cost (Le Houérou, 1976), even with very low rainfall of the order of 60 to 100 mm.

Importance of Desertization

Desertization affects huge areas in North Africa (Quezel, 1958; Le Houérou, 1959, 1962, 1968, 1969, 1973, 1975; Flohn and Kettata, 1971; Floret and Le Floch, 1973; Ionesco, 1972; Rapp, 1974; Wagner, 1974; Novikoff, 1974). Desertization is thought to affect several tens of thousands of hectares every year north of the Sahara (Le Houérou, 1968). In the Near and Middle East, the situation is exactly similar (Pabot, 1962, 1965, 1967; Pearse, 1970, 1971; Emberger, 1956, 1957; Reifenberg, 1953; Evenari, Shanan, and Tadmore, 1971; Zohary, 1963).

A similar situation occurs in the Sahel, where the recent drought (1968-73) has resulted in a considerable literature. Data from satellite imagery as well as field surveys suggest that desertization has also affected tens of thousands of hectares in the Sahel in recent years (Stebbing, 1938; Depierre and Gillet, 1971; Delwaulle, 1973; Boudet, 1972, 1974; Aubreville, 1973; MacLeod, 1976; Rapp, 1974; Gillet, 1974; Bille, 1973; Bille et al., 1972; Poupon, 1973; Michon, 1973; Glantz and Parton, 1976). According to the Sudan National Council for Research (1974), what was a semidesert region (100 to 300 mm) in that country between 14°N and 16°N and occupying 350,000 sq km, is now desert. The situation is similar in East Africa, especially in Somalia (Lewis, 1975), in Ethiopia, and in northern Kenya, as well as in the Kalahari. Desertization processes with very original features have also been described in southern Argentina (Movia, no date).

Desertization has become such a serious problem that several countries have set up large-scale projects to struggle against it (for example, Tunisia, Iran, Sudan). Several international meetings have been devoted to the subject during the last few months; UNEP, FAO, and UNESCO have set up global programs and projects to cope with desertization (FAO program on Ecological Management of Arid and Semi-Arid Rangeland; UNESCO's MAB Project No. 3 on Grazing Land Ecosystems, subprojects ARIDMED and ARIDTROP). In high political spheres the subject has been considered serious enough to justify the convening of a World Conference to be held in 1977.

Causes of Desertization

Is desertization a phenomenon caused by temporary drought periods, is it due to a general climatic trend towards aridity, or is it the result of human activities such as man's degradation of the biological environments in arid zones? These are basic issues, since the answers will influence the whole strategy to be applied in order to overcome the catastrophe.

If we are at the beginning of a dry period that will worsen, as some climatologists claim, then the only possible strategy is to withdraw populations from the areas of desertization and to settle them elsewhere. If, on the contrary, the phenomenon is governed by a temporary climatic fluctuation, desertization will end with the dry fluctuation and there is little to worry about in the long run. All that has to be done is to relieve the drought-stricken populations, enabling them to pull through until the weather situation improves in a few months or a few years. However, if desertization is primarily due to the misuse and degradation of natural resources by man and his animals, the situation could and should be attacked through education, extension services, and ecologically sound development projects that can ensure sustained productivity of arid lands and the welfare of the people living in arid regions.

Is There a General, Long-Term Increase in Aridity?

There are two conflicting sets of opinions on the subject of trends toward aridity. One group of specialists in dynamic climatology (Lamb, 1972, 1973, 1974; Winstanley, 1973; Bryson, 1973; Leroux, 1973) is of the opinion that we are now at the beginning of a long drought period and that it is likely to worsen in the long run, a situation we should be prepared to face. Their theory is based upon the expansion of cold air from the polar zones, provoking a shift towards the equator of the major high-pressure

belts that limit the advance of equatorial moist air—
"intertropical convergence" (ITCZ) or the "Front Intertropical"
(FIT) in French—toward the tropics. As there is no indication of
any climatic change at the poles, there is no reason why the present
situation would change and consequently why the forecasts should
be bleak for arid zones, especially those in the intertropical belts.

The second category of specialists, more numerous by far—
climatologists, hydrologists, ecologists, geographers, paleontolo-
gists, geologists—asserts that statistical studies of available weather records (which in some places date back only 130 to 135 years and
often only 80 to 100 years) do not allow any conclusions to be
drawn on the possibility of a systematic long-term change in
rainfall (Monod, 1958; Dubief, 1963; Le Houérou, 1968, 1969;
Flohn and Ketatta, 1971; Roche, 1973; Bernard, 1974; Mattei and
Rikja, 1975; Boudet, 1972; Gillet, 1974; Maley, 1973; Chaperon,
1973; Sircoulon, 1974; Swift, 1973, 1974; Davy, 1973, 1974; Grove,
1972, 1973; Goudie, 1972; Bunting, 1975; Giraud, 1973; Rapp,
1974).

The two conflicting views are of a different nature. The first is a
theory, perfectly logical and rational, whereas the second refers to
facts rigorously checked with sometimes sophisticated statistical
methods. This suggests two remarks. First, because a theory is
logical and rational, it is not necessarily true. The history of
science has witnessed many cases of pleasing theories which have
turned out to be wrong. Secondly, one should never confuse
theories and facts, a distinction which, incidentally, is a rudimentary basis of any epistemology.

The facts are that during the span of instrumental records (80 to
130 years) there have been several periods of consecutive years of
drought followed by series of years with rains above the long-term
averages. The 1910-15 drought in the Sahel was more severe than
the 1968-73 one, whereas the 1944-48 period was roughly similar
(Roche, 1973). Likewise, in North Africa the drought periods of
1920-25 and 1945-50 were worse than the last one (1959-62).

The recurrence of such droughts is extremely irregular and
totally unpredictable, and one, therefore, cannot use the word
"cycle" in reference to them; all scientists agree on this fact.
Running averages, for instance, introduce pseudo-cycles linked to
the length of the running period and are known as "Slutsky's
effect" in statistics (Roche, 1973; Giraud, 1973). Consequently our
conclusion is that there is no evidence of any climatic deterioration
as a general explanation to the phenomenon of desertization. This,

of course, does not mean that climate has no effect. Obviously, most of the irreversible actions take place during the recurrent drought periods when vegetation is under heavier pressure and has little or no chance of regeneration. Thus, desertization takes place in a more or less discontinuous way during these prolonged droughts.

Man and Human Actions: Deterioration of the Ecological Equilibrium

The present time witnesses a demographic explosion which is by no means less acute in the arid zones than in other parts of developing countries. In North Africa and in the Near and Middle East, for example, population has multiplied sixfold since the beginning of this century. The present growth rate is between 3 and 3.5% per year in most of these countries, a doubling period of 20 to 23 years. South of the Sahara the growth is somewhat slower: 2.5 to 3.0% per year, a doubling period of 23 to 28 years.

The reason for this demographic explosion is well known: peace, medicine (antibiotics), national and international solidarity, slow increase, if any, in income and standard of living, and insufficient education. Some of the consequences of the ever-increasing population pressures are (a) overgrazing, (b) overcultivation, (c) burning of ligneous species for fuel, and (d) careless development projects which result in pasture destruction. "To the north of the Sahara it is man who creates the desert; climate is only a supporting factor" (Le Houérou, 1959).

Overgrazing

As population grows, the numbers of livestock tend to grow at about the same rate (Le Houérou, 1962, 1968, 1973) in order to meet the minimum subsistence level of about two to four standard stock units (SSU) per person[2] (in cases where no cereals are grown). In fact, in many areas the human/animal population ratio falls well below what is usually considered as the minimum subsistence level, such as, for instance in the Adrar n'Iforas mountains (Swift, 1972, 1975) or in the Agadez-Air region (Le Houérou, 1972).

At present, animal population densities are far beyond carrying capacity (Le Houérou, 1962-1975; Pabot, 1962-1967; Pearse, 1970-1972; Boudet, 1972-1974; Bernus, 1972-1975; Peyre de Fabregues,

1963-1975; Le Houérou, Delhaye, and Sarson, 1974; Le Houérou, Claudin, and Haywood, 1975).The result is a progressive reduction in vegetation cover and increased wind erosion, trampling, sealing, increased runoff, higher water tables, and salinity—all mechanisms which feed the desertization process.

Overcultivation

For the same reasons there is a progressive creeping of cultivation into the pastoral area. Subsistence needs are considered to be about 250 kg of cereals per person per year, including seed reserves; since yields do not increase—on the contrary, official statistics show a clear decrease in yields in most of the countries of Africa and the Near East (Le Houérou, 1973)—more people means that more land must be tilled. As a matter of fact, cropping now takes place in areas receiving as little as 150 mm of annual rainfall in North Africa and the Near East and 250 mm in the Sahel.[3] Under such conditions crop expectancy is very low: 100 to 200 kg per ha per year. This amounts to one crop every third or fourth year and to three to four failures in five years. This unproductive and highly risky cropping situation results in the destruction of native fodder species which are then replaced either by annual weeds having little forage value or by unpalatable species such as *Peganum harmala* in the Mediterranean and *Calotropis procera* in the tropics.

In Niger a decreee prohibiting cropping north of the 350-mm isohyet was made in 1962. It was never enforced, however. In Algeria, cultivation has been prohibited in some steppic areas since 1973.

As cultivable land becomes insufficient in area, fallow is reduced, soil fertility is no longer restored, and yields decrease, as shown in table 2.1 taken from an official document by the government of Sudan.

Soil left barren after cropping or after crop failure is subject to wind erosion. In sandy cultivated and fallow fields, erosion rates of 1.3 tn per ha per month, or 10 mm of topsoil removed, have been measured in southern Tunisia (Floret and Le Floch, 1972). Similar rates have been reported in the Sahelian part of Senegal (Aubert, private communication).

Table 2.1. Production of Kordofan Province, 1960/61 – 1972/73

Year	Groundnuts			Sesame		
	Total area cropped (hectares)	Total prod. (tons)	Prod. (feddan*)	Total area cropped	Total prod. (tons)	Prod. (feddan)
1960/61	184,000	73,000	0.400	112,000	38,000	0.384
1963/64	200,820	.58,777	0.297	299,200	30,719	0.106
1966/67	211,200	45,657	0.216	382,000	31,560	0.083
1969/70	300,000	69,728	0.232	450,000	43,268	0.097
1972/73	810,000	73,690	0.090	778,940	14,722	0.090

*feddan = 1.03 acre

Uprooting Woody Species

Minimal wood consumption for domestic uses is about one kg per person per day, and often it is more than three times that figure. Given the fact that the average woody biomass of a steppe in good condition, either in the Mediterranean or in the tropics, is between 500 and 1000 kg per ha, including the main roots, each person dependent on this type of fuel would destroy at least half a hectare every year if there were no regeneration. If we assume that there is some regeneration on the average of every second year each family of five persons would still destroy more than one ha of woody steppe per year. As over 100 million people depend on this type of fuel for their daily needs in the arid zones of Africa and the Near and Middle East, we can estimate a theoretical destruction of about 25 million ha per year. This is, of course, mere speculation but it shows the magnitude of a very serious cause of desertization which is often overlooked. Woodcutting is a very serious problem around almost all towns and cities of the Sahelian and Sudanian zones of Africa. Many studies—for such cities as Niamey, Ouagadougou, and Kano—have been devoted to the subject.

Unsound Land and Water Development Policies

In many countries of Africa and the Near and Middle East, range development has meant in the past and still means vaccination campaigns and the drilling of boreholes irrespective of grazing capacities. The result has always been the same: the multiplication of livestock numbers, and the total destruction of the pastures within a radius of 15 to 30 km around boreholes (Bernus, 1971). During the 1968-73 drought in the Sahel most of the stock died of hunger, not of thirst. In the most arid zones the end product is a "borehole-made-desert" strewn with carcasses all over the Sahel (as in 1972 around the borehole of In Wagar, south of In Gall, Niger, and many other unfortunate places).

Conclusions

One should differentiate between, on one hand, true climatic deserts which have always been deserts, at least during historic times, such as Tenere, Tanezruft, Libyco-Egyptian, Arabic Sinai, Atacama, Namib, and on the other hand, the man-made deserts which occur in arid zones.

It has been written, and archaeologic evidence supports the point, that true deserts have been so for at least 2000 years. For instance, Marco Polo, writing about the Iranian desert in the thirteenth century in his "Description of the World" states: "There are no dwellings, everything is desert and great drought . . . and there are no wild animals either; they would find nothing to eat; the earth is so dry and hard that grass could not grow" (quoted by Monod, 1973). One could quote many similar texts from Roman, Greek, and Arab sources about the Sahara and the Near East (Le Houérou, 1968, 1969).

Desertization is concerned with the creep of desert-like conditions into areas where they should not climatologically exist, thus reducing the size of the ecumene. This expansion of deserts along their edges is due only to human action: to the permanent and increasing pressure of man and his animals on fragile and unstable ecosystems and to the misuse of natural resources through careless management (or lack of management). Desertization is discontinuous in time and space. It varies in time with the recurrence of prolonged drought periods when pressures on ecosystems reach a paroxysm, whereas the phenomenon is more or less halted during series of abnormally rainy years. It varies in space with the densities of human and animal population, which, in turn, are linked to water availability. Paradoxically, those areas where water conditions are better are those which undergo the most acute desertization (Le Houérou, 1959).

The image that has sometimes been offered of desertization progressing along a lengthy front line at a more or less constant speed of x km per year is childish. It has been written, for example, that "between 1927 and 1961 the average southern shift in isohyets in the Sahel was 0.8 km per year and therefore in the year 2000 they will have moved 140 km to the south in respect to 1926. . . ." This is, in my opinion, one of the nicest pieces of climate fiction that one could probably find in the specialized literature. In fact, the rate of the advance of desertization is not known; there are only very coarse estimates. However, assessment is made quite easy now by comparing sets of aerial photographs taken several years apart and also by using satellite imagery (MacLeod, 1976; Rapp, 1974). This makes it possible to measure the scope and importance of the phenomenon and to monitor its development.

Several authors have stressed the possible consequences of desertization on climate fluctuations as a result of the huge amount of dust produced and released into the atmosphere—dust which travels thousands of miles away from the areas where it originated (Rapp, 1974; MacLeod, 1976; Lundholm, 1975). It has been suggested that dust clouds might reduce rain-producing clouds. It has also been hypothesized that reduction in organic decay of vegetation in the large drought-stricken areas of the Sahel and of East Africa would have reduced the availability of ice-forming biogenic nuclei, thus reducing the amount of precipitation and creating a self-catalyzing feedback mechanism of desertization (Schnell, 1974; Parton and Schnell, 1975). Other studies using simulation models suggest that an increase in albedo (reflectivity) due to a reduction of vegetation cover may also reduce rainfall and create a feedback mechanism reinforcing desertization (Charney et al., 1975).

I shall conclude by quoting two prominent scientists who drew—almost 20 years ago—what I consider to be the right conclusions concerning desertization.

> The consensus seems to be that as our knowledge of these regions (deserts adjacent to the Mediterranean) improves one finds that, since the beginning of historic times, no appreciable, significant, general change in climate has occurred (Monod, 1958).

Arid lands predominate in the countries of Near and
Middle East occupying most of their surface. Those
regions are more or less "deserts" today; but they are
"false deserts" having been gradually desertized by man
during centuries of frantic misuse. The present-day vegetation is not in equilibrium with climate, far from it.
Climate has much higher potentials than those shown in
the vegetation existing now in these countries. . . . It is
man who has created these Near Eastern deserts (Emberger, 1957).

Notes

1. The words "reg," "serir," and "hammada" are understood to
have the meanings proposed by Monod (1937).
2. One SSU = 10 sheep or goats = 2 cows = 1 camel.
3. Agronomists consider minimum desirable rainfall for regular
cropping to be 400 mm in both cases.

References

Adam, J., 1967. Evolution de la végétation dans les sous parcelles
    rotégées de l'UNESCO-IFAN à Atar. *Bulletin IFAN* (Dakar),
    Sér. A, 29(1): 96-106.
Aubreville, A., 1949. *Climat, Forêts et Désertification de l'Afrique
    Tropicale.* Paris: Soc. Edit. Geogr. Marit. et Col.
Aubreville, A., 1973. Rapport de la mission forestière anglo-
    française Nigeria-Niger (Déc. 1936-Fév. 1937). *Bois et Forêts
    des Tropiques* (Paris), 148: 3-26.
Bernard, E. A., 1974. Aspects Scientifiques et Institutionnels de la
    Lutte Contre la Sécheresse au Sahel. Bruxelles: Académie
    Royale des Sciences d'Outremer.
Bernus, E., 1971. Possibilities and limits of pastoral watering plans
    in the Nigerian Sahel. Cairo: FAO Seminar on Settlement of
    Nomads.
Bernus, E., and G. Savonnet, 1973. Les problèmes de la sécheresse
    dans l'Afrique de l'Ouest. *Présence Africaine* (Paris), 2e Sér.,
    88: 114-138.

Baumer, M., 1973. Pour une stratégie de développement dans les zones Sahéliennes et Soudano-Sahéliennes. Mimeo/St/Spec. New York: UNDP Sahel Office.

Baumer, M., 1974. Un futur possible pour les zones arides et semi-arides de la Méditerranée. *Options Méditerranéennes* (Paris), p. 23.

Bille, J. C., 1973. L'écosystème sahélien de Fété olé. Essai de bilan au niveau de la production nette annuelle. Paris: Multigr. ORSTOM.

Bille, J. C., M. Lepage, G. Morel, and H. Poupon, 1972. Recherches écologiques sur une savane sahélienne du Ferlo Septentrional, Sénégal. Présentation de la Région. *La Terre et La Vie* (Paris), 26(3): 332-350.

Boudet, G., 1972. Désertification de l'Afrique tropicale sèche. *Adansonia* (Paris), Sér. 2, 12(4): 505-524.

Boudet, G., 1974. Rapport sur la situation pastorale dans les pays du Sahel (mimeo FAO/UNEP/IEMVT). Réunion techn. sur l'aménagement écol. des zones arides et semi arides d'Afrique et du Moyen Orient. Rome: FAO.

Boudet, G., 1975. Les pâturages et l'élevage au Sahel. Notes Techniques du MAB: le Sahel: bases écologiques de l'aménagement. Paris: UNESCO, pp. 29-34.

Bryson, R. A., 1973. Climatic modification by air pollution II. The Sahelian effect. University of Wisconsin: Institute of Environmental Studies Report no. 9.

Bunting, A. H., M. D. Dennett, J. Elston, and J. R. Milford, 1975. Seasonal rainfall forecasting in West Africa. *Nature*, 253: 622-623.

Bunting, A. H., M. D. Dennett, J. Elston, and J. R. Milford, 1975. Climatic factors. In *The State of Knowledge Report on Tropical Pastures*. Paris: UNESCO.

Buring, P., H.D.J. Van Heemst, and G. J. Staring, 1975. Computation of the absolute maximum food production of the world. Wageningen: Agricultural University, Department of Tropical Soil Science.

Chaperon, P., 1973. Quatre années de sécheresse dans le Sahel. Mimeo ORSTOM (Dakar) and Colloque International sur le Désertification (Nouakchott, Mauritania).

Charney, J., et al., 1975. Drought in the Sahara: a biophysical feedback mechanism. *Science*, 187(4175): 434-435.

Davy, E. G., 1973. A Programme for Development of Applications of Meteorology in the Sudano-Sahelian Zone. Geneva: W.M.O.

Davy, E. G., 1974. Drought in West Africa. *WMO Bulletin* (Geneva), 22: 18-23.

Delhaye, R., H. N. Le Houérou, and M. Sarson, 1974. L'amélioration des pâturages et de l'élevage dans la Hodna (Algérie). AGS: DP/ALG 66509 Rapp. Techn. No. 2. Rome: FAO.

Delwaulle, J. C., 1973. Désertification de l'Afrique au Sud du Sahara. *Bois et Forêts des Tropiques* (Paris), 149: 3-20.

Depierre, D., and H. Gillet, 1971. Désertification de la zone sahélienne du Tchad. *Bois et Forêts des Tropiques* (Paris), 139: 3-25.

Dubief, J., 1963. Contribution au problème des changements du climat survenus au cours de la période couverte par les observations météorologiques faites dans la Nord de l'Afrique. Paris: UNESCO, *Arid Zone Res.* 20: 75-78.

Emberger, L., 1956. L'intérêt de la connaissance de la végétation pour la mise en valeur du sol. Première réunion group trav. sur les herb. et prod. fourr. au Proche-Orient. Cairo: FAO.

Emberger, L., 1957. Mission au Moyen Orient (Rapport sommaire). Mimeo NS/AZ/327. Paris: UNESCO.

Evenari, M., L. Shanan, N. H. Tadmor, and Y. Aharoni, 1961. Ancient agriculture in the Negev. *Science*, 133(3457): 976-996.

Evenari, M., L. Shanan, N. H. Tadmor, and Y. Aharoni, 1967. Rainfall patterns in the central Negev desert. *Israel Expl. Journal*, 17(3): 163-184.

Evenari, M., L. Shanan, N. H. Tadmor, and Y. Aharoni, 1971. *The Challenge of a Desert*. Cambridge, Mass.: Harvard University Press.

Floret, C., and E. Le Floch, 1973. Production, sensibilité et évolution de la végétation et du milieu en Tunisie présaharienne. Montpellier: CEPE, doc. no. 71.

Firouz, E., 1974. *Environment in Iran*. Teheran National Society for the Conservation of Natural Resources and Human Environment.

Firouz, E., 1975. Opening address to the regional meeting on de-desertization and arid lands ecology (multigraph). Teheran: Prime Ministry, Department of the Environment.

Flohn, H., and M. Kettata, 1971. Etude des conditions climatiques de l'avance du Sahara Tunisien. Note technique No. 116, Genève: OMM(WMO).

Franclet, A., and H. N. Le Houérou, 1971. Les Atriplex en Tunisie et en Afrique du Nord. Rome: FAO, Dep. Forêt.

Gillet, H., 1974. Tapis végétal et pâturage du Sahel. In *Notes Techniques du MAB: Le Sahel: bases écologiques de l'aménagement*. Paris: UNESCO, pp. 21-28.

Giraud, J. M., 1973. Y-a-t-il périodicité dans la répartition annuelle des précipitations? Mimeo Coll. Desertif. *Bulletin IFAN* (Dakar).

Giraud, J. M., 1973. Recherche des cycles dans les pluies annuelles de Dakar (1901-1972) et du Sénégal (1924-1972). Dakar: Multigr. ASECNA-Et. No. 31.

Glantz, M., and W. Parton, 1976. Weather and climate modification and the future of the Sahara. In *Politics of a Natural Disaster: Case of the Sahel Drought*. M. H. Glantz, ed. New York: Praeger Publishers.

Goudie, A. S., 1972. *The concept of post-glacial progressive desiccation*. Oxford: School of Geography.

Grove, A. T., 1971. *Africa South of the Sahara*, 2nd edition. London: Oxford University Press.

Grove, A. T., 1972. Climate change in Africa in the last 20,000 years. In *Les problèmes du développement du Sahara Septentrional Coll.* (Ouargla, 1971), Alger: Vol. 2, Instit. Geogr.

Grove, A. T., 1973. Desertification in the African environment. In *Drought in Africa*, D. Dalby and R. Harrison Church, eds. London: University of London School of Oriental and African Studies (henceforth SOAS), pp. 33-45.

Hartley, B. J., 1972. Somalia: Survey of northern rangelands. Report on project results, conclusions and recommendations. Mimeo UNDP/SF/SOM, pp. 70-512. Rome: FAO.

Ionesco, T., 1972. Pastoralisme et désertisation. Sem. Nation. Désertisation (Gabès) Dec. 1972. Tunis: Inst. Nal. Rech. Agron.

Kassas, M., 1970. Desertification versus potential for recovery in circum-Saharan territories. In *Arid Lands in Transition*, H. Dregne, ed. Washington: AAAS. Pp. 123-39.

Knowles, G., 1954. Scald reclamation in the Hay district. *Journ. Soil. Cons. Serv. New South Wales* (Sydney), 10(3): 149-156.

Lamb, H. H., 1966. *The Changing Climate*. London: Methuen and Co.

Lamb, H. H., 1966. Climate in the 1960s. *Geogr. J.*, 132: 183-212.

Lamb, H. H., 1972. *Climate: Past, Present and Future* (Vol. 1: Fundamentals and Climate Now). London: Methuen and Co.

Lamb, H. H., 1973. Some comments on atmospheric pressure variations in the northern hemisphere. In *Drought in Africa*. London: SOAS.

Lamb, H. H., 1974. Drifting towards drought. *Geogr. Mag.*, 46: 455.

Le Floch, E., and C. Floret, 1972. Désertisation, dégradation et régéneration de la végétation pastorale de la Tunisie présaharienne. Semin. Nat. Désertisation (Gabès) Dec. 1972. Tunis: Inst. Nal Rech. Agron.

Le Houérou, H. N., 1958. L'élevage dans le sud Tunisien et ses perspectives en fonction de la valeur des parcours. *Terre de Tunisie* (Tunis).

Le Houérou, H. N., 1959. *Recherches écologiques et floristiques sur la végétation de la Tunisie méridionale*. Alger: Inst. Rech. Sah.

Le Houérou, H. N., 1962. *Les pâturages naturels de la Tunisie aride et désertique*. Paris: Inst. Sces. Econ. Appl. Tunis.

Le Houérou, H. N., 1965. Improvement of Natural Pastures and Fodder Resources: Report to the Govt. of Libya (EPTA Report). Rome: FAO.

Le Houérou, H. N., 1968. La désertisation du Sahara septentrional et des steppes limitrophes. *Ann. Alger. de Géogr.* (Alger), 6: 2-27.

Le Houérou, H. N., 1969a. Principes méthodes et techniques d'amélioration pastorale et fourragère. Tunisie. Rome: FAO.

Le Houérou, H. N., 1969b. La végétation de la Tunisie steppique (avec référence aux végétations analogues du Maroc, d'Algérie et de Libye). Ann. Inst. Nat. Rech. Agr. (Tunis) 42: 5.

Le Houérou, H. N., 1970. North Africa past present future. In *Arid Lands in Transition*, H. Dregne, ed. Washington: AAAS, pp. 227-278.

Le Houérou, H. N., 1971. An assessment of the primary and secondary production of the arid grazing land ecosystems of North Africa. Leningrad: International Symp. Found. of Ecosystem Product in Arid Zone, summary pp. 168-172.

Le Houérou, H. N., 1973a. Peut-on lutter contre la désertisation? Coll. Desertif. *Bulletin IFAN* (Dakar).

Le Houérou, H. N., 1973b. Ecologie, démographie et production agricole dans les pays méditerranéens du Tiers Monde. *Options Medit.* (Paris), 17: 53-61.

Le Houérou, H. N., 1974. Deterioration of the ecological equilibrium in the arid zones of North Africa. Bet Dagan, Israel. Volcani Institute for Agricultural Research. Spec. Publ. No. 39, pp. 54-57.

Le Houérou, H. N., 1975. Report on a consultation mission to the Range Organization of Iran. AGPC/Misc. No. 33. Rome: FAO.

Le Houérou, H. N., 1976. Ecological management of arid grazing land ecosystems. In *Politics of a Natural Disaster: The Case of the Sahel Drought*, M. H. Glantz, ed. New York: Praeger Publishers.

Lewis, I. M., 1975. *Abaar: the Somali Drought*. London: International African Institute.

Leroux, M., 1973. Climatologie dynamique de l'Afrique à paraître. In Trav. et Doc. Geog. Trop.—C.E.G.E.T.

Leroux, M., 1975. La circulation générale de l'atmosphère et les oscillations climatiques tropicales. Coll. Intern. Desert, *Bulletin IFAN* (Dakar).

Lundholm, B. G., 1971. Remote sensing and global environment monitoring. *Proceed. 7th Intern. Symp. Rem. Sens. Env.*, pp. 153-159.

Lundholm, B. G., 1975. Environmental monitoring needs in Africa. In *African Environment: Special Report #1*, Paul Richards, ed. London: International African Institute, pp. 9-24.

Maley, J., 1973. Un nouveau mechanisme des changements climatiques aux basses latitudes. *Ass. Sénégal Et. Quatern. Ouest Afr. Bull. Liaison* (Dakar), No. 37-38, pp. 31-40.

Malcolm, C. V., 1972. Establishing shrubs in saline environment. In *Wildland Shrubs, Their Biology and Utilization*. Ogden, Utah: USDA Forest Serv. Techn. Dept. INT 1, pp. 392-403.

MacLeod, N. H., 1976. Dust in the Sahel: causes of drought? In *Politics of a Natural Disaster: The Case of the Sahel Drought*, M. H. Glantz, ed. New York: Praeger Publishers.

Mattei, F., and D. Rijks, 1975. Problèmes agricoles et de l'élevage dans la région Soudano-Sahélienne. Rome: FAO.

Michon, P., 1973. Le Sahara avance-t-il vers le sud? *Bois et Forêts des Tropiques* (Paris), pp. 69-80.

Monod, T., 1937. *Meharées, Explorations au Vrai Sahara*. Paris: Edit. Jesers.

Monod, T., 1951. Note sur un projet d'établissement au Sahara Sud Occidental de parcelles expérimentales protégées. IFAN, Dakar.

Monod, T., 1954. Mode contracté et diffus de la végétation Saharienne. In *Biology of Deserts*, J. L. Cloudsley-Thompson, ed.

London: Inst. of Biology, pp. 35-44.

Monod, T., 1957. Les grandes divisions chorologiques de l'Afrique. London: Cons. Scient. Afr. au Sud du Sahara, Publ. No. 24.

Monod, T., 1958. Parts respectives de l'homme et des phenomènes naturels dans la dégradation des paysages et le déclin des civilisations à travers le Monde Méditerranéen 1.S., avec les déserts et semi-déserts adjacents, au cours des derniers millenaires. Athens: UICN, 73 Reun. Tech.

Monod, T., 1973a. Le dégradation du monde vivant: flore et faune. Coll. Desertif. *Bulletin IFAN* (Dakar).

Monod, T., 1973b. *Les Déserts*. Paris: Horizons de France Edit.

Monod, T., ed., 1975. Pastoralism in Tropical Africa. *Proceed. of the Niamey Symposium (Dec. 1972)*. London: Oxford University Press.

Movia, C. (no date). Estudio sobre erosión en la zona de Río Mayo, Chubut (Argentina). Unpublished mimeo.

Naegele, A., 1959. La végétation de la zone aride, les parcelles protégées d'Atar. *La Nature* (Paris), 3286: 72-76.

Nichaeva, N. T., 1969. Pasture of arid lands of the USSR and their use in arid lands in a changing world. *Abstr. proceed. Intern. Conf. AAAS and University of Arizona*, Tucson.

Nichaeva, N. T., and S. Y. Prikhod'ko, 1968. Sown winter ranges in the foothill desert of Soviet central Asia. Jerusalem: Israel Progr. for Scient. Transl.

Niknam, F., and Ahranjani, 1975. Dunes and development in Iran. Teheran: Ministry of Agriculture and Nat. Res., Forestry and Range Organiz.

Novikoff, G., F. Wagner, and A. Skouri, 1973-75. Tunisian presaharan progress reports. Nos. 1, 2, 3. US/IBP Desert Biome. Logan: Utah State University.

Novikoff, G., 1974. The desertisation of rangelands and cereal cultivated lands in presaharan Tunisia. A statement of some possible methods of control. Stockholm: SIES Report no. 3.

Pabot, H., 1962. Comment briser le cercle vicieux de la désertification dans les regions sèches d'Orient. Rome: FAO.

Pabot, H., 1965. L'Action de l'homme sur la végétation naturelle et ses conséquences dans les pays secs ou arides du Proche et Moyen Orient. Rome: FAO, Div. Prod. Veget.

Pabot, H., 1967. Pasture development and range improvement through botanical and ecological studies. Report to the Government of Iran. Rome: FAO, EPTA report no. 2311, PL Div.

Parton, W. J., and R. C. Schnell, 1975. Use of an ecosystem model to study potential interactions between the atmosphere and the biosphere. San Francisco: Summer Simulation Conference.

Pearse, C. K., 1970. Range deterioration in the Middle East. Surfers Paradise, Queensland: *Proceed. XI Intern. Grass. Congr.*, pp. 26-30.

Pearse, C. K., 1971. Grazing in the Middle East—past and present and future. *Journ. Range Management*, 24(1): 13-16.

Poupon, H., 1973. Influence de la sécheresse 1972-73 sur la végétation d'une savane sahélienne du Ferlo Septentrional (Sénégal). Coll. Desertif. *Bulletin IFAN* (Dakar).

Quezel, P., 1958. Quelques aspects de la dégradation du paysage végétal au Sahara et en Afrique du Nord. Athens: UICN, 7e Reun. Tech.

Rapp, A., 1974. A review of desertization in Africa: water, vegetation and man. Stockholm: Secretariat for International Ecology.

Reifenberg, A., 1953. The struggle between the desert and the sown. *Proceed. Intern. Symp. on Desert Res.* Jerusalem: Res. Council Israel, pp. 378-391.

Roche, M., 1973. Note sur la sécheresse en Afrique de l'Ouest. In *Drought in Africa*. London: SOAS, pp. 53-61.

Roche, M., 1973. Incidences climatiques et hydrologiques de la sécheresse. In *Technique et Développement* (Paris), 10: 4-15.

Shaidaee, G., 1974. Conservation, rehabilitation and improvement of rangelands in Iran. Teheran: Ministry of Agriculture and Nat. Res., Forestry and Range Organiz.

Shaidaee, G., and F. Niknam, 1974. Resources conservation in Iran. Teheran: Ministry of Agriculture and Nat. Res., Forestry and Range Organiz.

Sherbrooke, W. C., and P. Paylore, 1973. *World Desertification: Causes and Effect—A Literature Review and Annotated Bibliography*. Arid Land Inf. Paper no. 3. Tucson: University of Arizona Office of Arid Land Studies.

Sircoulon, J., 1974. Les données climatiques et hydrologiques de la sécheresse en Africa de l'Ouest sahélienne. ORSTROM (Paris) and SIES (Stockholm).

Schnell, R. C., 1974. Biogenic and inorganic sources for ice nuclei in the drought stricken areas of the Sahel. Boulder, Colo.: National Center for Atmospheric Research.

Schnell, R. C., 1974. Modelling the climate. *Science*, 21 September 1974.

Stebbing, E. P., 1938. The man-made desert in Africa. *Journ. Roy. Afr. Soc.* 37 (supplement).

Stebbing, E. P., 1953. *The Creeping of the Desert in the Sudan and Elsewhere in Africa.* Khartoum: McCorquale.

Sudan, Republic of, 1974. Desert encroachment control. Khartoum: National Council for Research.

Swift, J., 1973. Disaster and a Sahelian nomad economy. In *Drought in Africa*, D. Dalby and R. Harrison Church, eds. London, SOAS.

Swift, J., 1974. The future of Twareg pastoral nomadism in the Malian Sahel. Cambridge: Sympos. on the future of traditional societies.

Swift, 1974. The cause of the Sahel disaster. University of Sussex: Inst. for the Study of International Organisation.

Swift, J., 1975. Development options for the Sahel. University of Sussex: Inst. for the Study of International Organisation.

Swift, J., 1975. Pastoral nomadism as a form of land-use: the Tuareg of the Adra n'Iforas. In *Pastoralism in Tropical Africa*, T. Monod, ed. London: Oxford University Press. Pp. 443-454.

U.S. AID, 1972. Desert encroachment on arable lands: significance, causes and control. Washington: U.S. A.I.D. Office of Science and Technology.

Wagner, F. H., 1974. Consideration of some causes and solutions to the problem of desertization. Rep. U.S., MAB 3. Logan: Utah State University.

Winstanley, D., 1973. Rainfall patterns and general atmospheric circulation. *Nature* (London), 245: 190-194.

Winstanley, D., 1973. Recent rainfall trends in Africa, the Middle East and India. *Nature* (London), 243: 464-465.

Winstanley, D., 1974. Seasonal rainfall forecasting in West Africa. *Nature* (London), 248: 464.

Wickens, G. E., 1962. Savanna development in the Sudan: plant ecology. Rome: FAO, Prod. Div.

Zohary, M., 1963. On the geobotanical structure of Iran. Jerusalem: *Bulletin of the Research Council of Israel.*

Zumer-Linder, M. Ecological dilemma of agricultural ecosystems in the tropics. Stockholm: SIES.

Zumer-Linder, M. Some comments and illustration to the "Village Forestry." Stockholm: SIES.

3

The Other Energy Crisis

Erik P. Eckholm

"Even if we somehow grow enough food for our people in
the year 2000, how in the world will they cook it?"
— *an Indian official, 1975*

Dwindling reserves of petroleum and artful tampering with its
distribution are the stuff of which headlines are made. Yet for more
than a third of the world's people, the real energy crisis is a daily
scramble to find the wood they need to cook dinner. Their search
for wood, once a simple chore and now, as forests recede, a day's
labor in some places, has been strangely neglected by diplomats,
economists, and the media. But the firewood crisis will be making
news—one way or another—for the rest of the century.

While chemists devise ever more sophisticated uses for wood,
including cellophane and rayon, at least half of all the timber cut
in the world still fulfills its original role for humans—as fuel for
cooking and, in colder mountain regions, a source of warmth.
Nine-tenths of the people in most poor countries today depend on
firewood as their chief source of fuel. And all too often, the growth
in human population is outpacing the growth of new trees—not

This chapter is adapted from Eckholm's book *Losing Ground: Environmental Stress and World Food Prospects* (New York: W. W. Norton & Co.,
1976). It is a slightly revised *Worldwatch Paper 1* (Washington, D.C.:
Worldwatch Institute, September 1975).

surprising when the average user burns as much as a ton of firewood a year (Openshaw, 1974; UN/FAO, 1967). The results are soaring wood prices, a growing drain on incomes and physical energies in order to satisfy basic fuel needs, a costly diversion of animal manures for cooking food rather than producing it, and an ecologically disastrous spread of treeless landscapes.

The firewood crisis is probably most acute today in the countries of the densely populated Indian subcontinent, and in the semiarid stretches of central Africa fringing the Sahara Desert, though it plagues many other regions as well. In Latin America, for example, the scarcity of wood and charcoal is a problem throughout most of the Andean region, Central America, and the Caribbean.

An Economic Burden

As firewood prices rise, so does the economic burden on the urban poor. One typical morning on the outskirts of Kathmandu, Nepal's capital city, I watched a steady flow of people—men, women, children, and the very old—trudge into the city with heavy, neatly chopped and stacked loads of wood on their backs (figure 3.1). I asked my taxi driver how much their loads, for which they had walked several hours into the surrounding hills, would sell for. "Oh wood, a very expensive item!" he exclaimed without hesitation. Wood prices are a primary topic of conversation in Kathmandu these days. "That load costs 20 rupees now. Two years ago it sold for six or seven rupees." This 300 percent rise in the price of fuel wood has in part been prompted by the escalating cost of imported kerosene, the principal alternative energy source for the poor. But firewood prices have risen much *faster* than kerosene prices, also reflecting the growing difficulty with which wood is procured. It now costs as much to run a Kathmandu household on wood as on kerosene.

The costs of firewood and charcoal are climbing throughout most of Asia, Africa, and Latin America. Those who can, pay the price, and thus must forego consumption of other essential goods. Wood is simply accepted as one of the major expenses of living. In Niamey, Niger, deep in the drought-plagued Sahel in West Africa, the average manual laborer's family now spends nearly one-fourth

Figure 3.1 Colombian boy gathering firewood in the Andes. Credit: Carl Purcell, U.S. Agency for International Development.

of its income on firewood. In Ouagadougou, Upper Volta, the portion is 20 to 30% (Delwaulle, 1973; DuBois, 1974). Those who can't pay so much may send their children, or hike themselves, out into the surrounding countryside to forage—if enough trees are within a reasonable walking distance. Otherwise, they may scrounge about the town for twigs, garbage, or anything burnable.

In some Pakistani towns now, people strip bark off the trees that line the streets; thus meeting today's undeniable needs can impoverish the future. When I visited the Chief Conservator of Forests in Pakistan's North West Frontier Province at his headquarters in the town of Peshawar, he spoke in a somewhat resigned tone of stopping his car the previous day to prevent a woman from pulling bark off a tree. "I told her that peeling the bark off a tree is just like peeling the skin off a man," he said. Of course the woman stopped, intimidated by what may have been the most personal encounter with a senior civil servant she will have in her lifetime, but she doubtless resumed her practice shortly, for what else, as the Chief Conservator himself asked, was she to do?

It is not in the cities but in rural villages that most people in the affected countries live, and where most firewood is burned. The rural, landless poor in parts of India and Pakistan are now facing a new squeeze on their meager incomes. Until now they have generally been able to gather wood for free among the trees scattered through farmlands, but as wood prices in the towns rise, landlords naturally see an advantage in carting available timber into the nearest town to sell rather than giving it to the nearby laborers. While this commercialization of firewood raises the hope that entrepreneurs will see an advantage in planting trees to develop a sustainable, labor-intensive business, so far a depletion of woodlands has been the more common result. And the rural poor, with little or no cash to spare, are in deep trouble in either case.

With the farmland trees and the scrubby woodlands of unfarmed areas being depleted by these pressures, both the needy and the entrepreneurs are forced to poach for fuel wood in the legally protected, ecologically and economically essential national forest reserves. The gravity of the poaching problem in India has been reflected in the formation of special mobile guard-squads and mobile courts to try captured offenders, but law enforcement measures have little effect in such an untenable situation. Acute firewood scarcity has undermined administrative control even in China, where trees on commune plantations are sometimes surreptitiously uprooted for fuel almost as soon as they are planted (Richardson, 1966; see also Government of India, 1973).

Trees are becoming scarce in the most unlikely places. In some of the most remote villages in the world, deep in the once heavily forested Himalayan foothills of Nepal, journeying out to gather firewood and fodder is now an *entire day's* task. Just one generation ago the same expedition required no more than an hour or two (Bishop, 1971).

Ecological Consequences

Because those directly suffering its consequences are mostly illiterate, and wood shortages lack the photogenic visibility of famine, the firewood crisis has not provoked much world attention. And in a way there is little point in calling this a world problem, for fuel-wood scarcity, unlike oil scarcity, is always localized in its apparent dimensions. Economics seldom permit fuel wood to be carried or trucked more than a few hundred miles from where it grows, let alone the many thousands of miles traversed by the modern barrel of oil. To say that firewood is scarce in Mali or Nepal is of no immediate consequence to the Boy Scout building a campfire in Pennsylvania, whereas his parents have already learned that decisions in Saudi Arabia can keep the family car in the garage.

Unfortunately, however, the consequences of firewood scarcity are seldom limited to the economic burden placed on the poor of a particular locality. The accelerating degradation of woodlands throughout Africa, Asia, and Latin America, caused in part by fuel gathering, lies at the heart of what will likely by the most profound ecological challenge of the late twentieth century. On a global basis, an ecological threat to human well-being far more insidious and intractable than the industrial pollution of our air and water—which has preempted thinking on environmental quality—is the undermining of the productivity of the land itself through soil erosion, increasingly severe flooding, creeping deserts, and declining soil fertility. All these problems are accentuated by deforestation, which is spreading as lands are cleared for agriculture and as rising populations continue their search for firewood. Rainwater falling on tree-covered land tends to soak into the ground rather than rush off; erosion and flooding are thus reduced, and more water seeps into valuable underground pools and spring sources.

The Dust Bowl years in the Great Plains of the thirties taught Americans the perils of devegetating a region prone to droughts. The images provided by John Steinbeck in *The Grapes of Wrath* of the human dislocation wrought by that interaction of man, land, and climate could easily describe present-day events in large semiarid stretches of Africa along the northern and southern edges of the Sahara, and around the huge Rajasthan Desert in northwest India. Overgrazing by oversized herds of cattle, goats, and sheep is the chief culprit, but fuel-wood gathering is also an important contributor to the destruction of trees in these regions. Firewood is a scarce and expensive item throughout the sub-Saharan fringe of Africa, all the way from Senegal to Ethiopia, but citizens in towns like Niamey are paying a much higher price than they realize for their cooking fuel. The caravans that bring in this precious resource are contributing to the creation of desert-like conditions in a wide band below the desert's edge. Virtually all trees within 70 km of Ouagadougou have been consumed as fuel by the city's inhabitants, and the circle of land "strip-mined" for firewood—without reclamation—is continually expanding.

Similar pressures of overgrazing and deforestation in North Africa are having the same consequences. H. N. Le Houérou (1970) of the International Livestock Center for Africa (ILCA) in Addis Ababa, Ethiopia, figures that 100,000 ha of land are lost to the desert *each year* due to human activities in Algeria, Morocco, Libya, and Tunisia. In an interesting experiment, the Libyans and others have tried spraying oil on the sand dunes to hold them back, but most arid countries will have to follow Algeria's recent lead in undertaking a massive tree-planting campaign if they hope to stop the sand. India, too, is forfeiting farmlands and rangelands to desert sands, while vaster dry regions, which stretch eastward from the Rajasthan Desert and constitute perhaps a fifth of the country, now present a nearly treeless landscape. Wind erosion is chronic and the agricultural yields extracted from the infertile subsoil are either falling or have finally stabilized at a low level.

Dangerous Substitutes

In the Indian subcontinent, the most pernicious result of fire-wood scarcity is probably not the destruction of tree cover itself, but the alternative to which a good share of the people in India, Pakistan, and Bangladesh have been forced. A visitor to almost any village in the subcontinent is greeted by omnipresent pyramids of hand-molded dung patties drying in the sun. In many areas these dung cakes have been the only source of fuel for generations, but now, by necessity, their use is spreading further. Between 300 and 400 million tn of wet dung—which shrinks to 60 to 80 million tn when dried—is annually burned for fuel in India alone, robbing farmland of badly needed nutrients and organic matter. The plant nutrients wasted annually in this fashion in India equal more than a third of the country's chemical fertilizer use. Looking only at this direct economic cost, it is easy to see why the country's National Commission on Agriculture (Government of India, 1973) recently declared that "the use of cow dung as a source of non-commercial fuel is virtually a crime." Dung is also burned for fuel in parts of the Sahelian zone in Africa, Ethiopia, Iraq, and in the nearly treeless Andean valleys and slopes of Bolivia and Peru, where the dung of llamas has been the chief fuel in some areas since the days of the Incas (Adeyoju and Enabor, 1973; U.S. Department of the Interior, 1964; Brown, 1973; UNDP/FAO, 1973; República del Peru, 1974; Mason, 1957).

Even more important than the loss of agricultural nutrients is the damage done to soil structure and quality through the failure to return manures to the fields. Organic materials—humus and soil organisms which live in it—play an essential role in preserving the soil structure and fertility needed for productive farming. Organic matter holds the soil in place when rain falls and wind blows, and reduces the wasteful, polluting runoff of chemical nutrients where they are applied, thus increasing the efficiency of their use. These considerations apply especially to the soils in tropical regions where most dung is now burned, because tropical topsoils are usually thin and, once exposed to the harsh treatment of the burning sun and torrential monsoon rains, are exceptionally prone to erosion, and to losing their structure and fertility.

Peasants in the uplands of South Korea have found another, equally destructive way to cope with the timber shortage. A United Nations forestry team visiting the country in the late 1960s found not only live tree branches, shrubs, seedlings, and grasses being cut for fuel; many hillsides were raked clean of all leaves, litter, and burnable materials. Raking in this fashion, to meet needs for home fuel and farm compost, robs the soil of both a protective cover and organic matter, and the practice was cited by the U.N. experts as "one of the principal causes of soil erosion in Korea" (UN-DP/FAO, 1969). Firewood scarcity similarly impairs productivity

in Eastern Nigeria, where the Tiv people have been forced to
uproot crop residues after the harvest for use as fuel (Vermeer,
1970). Traditionally, the dead stalks and leaves have been left to
enrich the soil and hold down erosion.

A Closing Circle

The increasing time required to gather firewood in many
mountain villages of Nepal is leading to what the kingdom's
agricultural officials fear most of all. For, once procuring wood
takes too long to be worth the trouble, some farmers start to use cow
dung, which was formerly applied with great care to the fields, as
cooking fuel. As this departure from tradition spreads, the fertility
of the hills, already declining due to soil erosion, will fall sharply.
In the more inaccessible spots there is no economic possibility
whatsoever of replacing the manure with chemical fertilizers.

And so the circle starts to close in Nepal, a circle long completed
in parts of India. As wood scarcity forces farmers to burn more
dung for fuel, and to apply less to their fields, falling food output
will necessitate the clearing of ever larger, ever steeper tracts of
forest—intensifying the erosion and landslide hazards. Even Nepal's key economic planning body, the National Planning Commission, now says that if present trends continue, a "semi-desert
type of ecology in the hilly regions" will be created.

Though most dramatically apparent in Nepal, this same cluster
of phenomena threatens the future habitability of the entire stretch
of Himalayan foothills, from Afghanistan through northern
Pakistan, India, and Nepal to Burma. And the negative consequences by no means stop at the base of the hills. When soil washes away
it must relocate somewhere, and the rising load of silt carried by
Asia's rivers is choking up expensive reservoirs and irrigation
works. Most threatening of all to food production prospects on the
Indian subcontinent, where nearly one in every five human beings
lives, is a rise in the frequency and severity of flooding in Pakistan,
India, and possibly Bangladesh, the result of denuded watersheds
off which rainfall rushes quickly, and of the excessive load of
sediment from upstream that builds up river beds, reducing their
capacity to channel water.

Firewood scarcity, then, is intimately linked to the food problem
facing many countries in two ways. Deforestation and the diversion of manures to use as fuel are sabotaging the land's ability to

produce food. Meanwhile, as an Indian official put it, "Even if we somehow grow enough food for our people in the year 2000, how in the world will they cook it?"

B. B. Vohra, a senior Indian agricultural official who has pushed his government ahead on numerous ecological causes, shook his head as we talked in his New Delhi office. "I'm afraid that we are approaching the point of no return with our resource base. If we can't soon build some dramatic momentum in our reforestation and soil conservation programs, we'll find ourselves in a down-ward spiral with an irresistible momentum of its own." Without a rapid reversal of prevailing trends, in fact, India will find itself with a billion people to support and a countryside that is little more than a moonscape. But the politicians, in India and other poor countries, will soon start taking notice, for they will begin to realize that if the people can't find any firewood, they will surely find something else to burn.

A Renewable Resource

The firewood crisis is in some ways more, and in others less, intractable than the energy crisis of the industrialized world. Resource scarcity can usually be attacked from either end, through the conservation of demand or the expansion of supply. The world contraction in demand for oil in 1974 and early 1975, for example, helped to ease temporarily the conditions of shortage.

The poor, like the rich, are faced with the necessity of energy conservation. Millions of families cook over wood or charcoal stoves that are extremely inefficient in the use of heat. The adoption of simple wood stoves that waste less heat could substantially reduce per capita firewood needs.

Even if more efficient stoves are disseminated, however, future firewood needs in the developing countries will, in the absence of alternative energy sources, be influenced largely by population growth.

The simple arithmetic of exponential growth suggests the immensity of the fuel-wood challenge facing many poor countries. As table 3.1 shows, 99% of Tanzania's population burns an average of 1.8 tn of wood per person each year. Since Tanzania's population is now just over 15 million, about 27 million tn of wood will be burned for fuel in 1975. But if Tanzania's population should continue growing at the present rate of 2.7% annually, it

Table 3.1. Fuel Wood Consumption in Tanzania, Gambia, and Thailand

	Fuel Wood Consumption Per Capita (tons/year)	Fuel Wood as Share of Total Timber Consumption (percent)	Fuel Wood Users as Share of Total Population (percent)
Tanzania	1.8	96	99
Gambia	1.2	94	99
Thailand	1.1	76	97

Source: Adapted from Openshaw, 1974.

will multiply *14-fold in a century.* If the proportion of wood users should remain constant, the consumption of wood for fuel would also multiply 14-fold to reach 370 million tn per year.

Per capita wood use in Thailand, at 1.1 tn per year, is somewhat lower than that in Tanzania. Yet Thailand's current population growth rate will produce a 16-fold increase in its population of 42 million in 100 years, resulting in a total of 672 million people—more than now live in India. It is difficult to envisage how Thailand's rich tropical forests could survive pressures of this magnitude. Even if the demographers are surprised by quick progress in slowing population growth over the next few decades, the demand for basic resources like firewood will still push many countries to their limits.

Fortunately trees, unlike oil, are a renewable resource when properly managed. The logical immediate response to the firewood shortage, one that will have many incidental ecological benefits, is to plant more trees in plantations, on farms, along roads, in shelter belts, and on unused land throughout the rural areas of the poor countries. For many regions, fast-growing tree varieties are available that can be culled for firewood inside of a decade.

The concept is simple, but its implementation is not. Governments in nearly all the wood-short countries have had tree-planting programs for some time—for several decades in some cases. National forestry departments in particular have often been aware of the need to boost the supply of wood products and the need to preserve forests for a habitable environment. But several problems have plagued these programs from the beginning.

One is the sheer magnitude of the need for wood, and the scale of the growth in demand. Population growth, which surprised many with its acceleration in the post-war era, has swallowed the moderate tree-planting efforts of many countries, rendering their impact almost negligible. Wood-producing programs will have to be undertaken on a far greater scale than most governments presently conceive if a real dent is to be made in the problem.

The problem of scale is closely linked to a second major obstacle to meeting this crisis: the perennial question of political priorities and decision-making time frames. What with elections to win, wars to fight, dams to build, and hungry mouths to feed, it is hard for any politician to concentrate funds and attention on a problem so diffuse and seemingly long-term in nature. Some ecologists in the poor countries have been warning their governments for decades about the dangers of deforestation and fuel shortages, but tree-planting programs don't win elections. This phenomenon is of course quite familiar to all countries, not just the poorest. In the United States, resource specialists pointed out the coming energy crisis throughout the 1960s, but it took a smash in the face in 1973 to wake up the government, and as of 1977 the country still can hardly be said to have tackled the energy challenge head on.

Despite these inherent political problems, India's foresters made a major breakthrough a few years back as the government drew up its five-year development plan for the mid- to late seventies. Plans were laid for the large-scale establishment of fast-growing tree plantations, and for planting trees on farms and village properties throughout the country (Government of India, 1973). A program is going ahead now, but there have been some unexpected events since the projects were first contemplated several years ago: the quintupling of the world price of petroleum, the tripling in price and world shortages of grains and fertilizers, a wholesale diversion of development funds just to muddle through 1974 without a major famine and a total economic breakdown. India's development efforts were set back several years by recent events, and forestry programs have not been immune to this trend.

Political, Cultural, and Administrative Tangles

Even when the political will is there and the funds are allocated, implementing a large-scale reforestation campaign is an unexpectedly complex and difficult process. Planting millions of trees and successfully nurturing them to maturity is not a technical, clearly boundaried task like building a dam or a chemical fertilizer plant. Tree-planting projects almost always become deeply enmeshed in the political, cultural, and administrative tangles of a rural locality; they touch upon, and are influenced by, the daily living habits of many people, and they frequently end in failure.

Most of the regions with too few trees also have too many cattle, sheep, and goats. Where rangelands are badly overgrazed, the leaves of a young sapling present an appetizing temptation to a foraging animal. Even if he keeps careful control of his own livestock, a herdsman may reason that if his animals don't eat the leaves, someone else's will. Marauding livestock are prime destroyers of tree-planting projects throughout the less developed world. Even if a village is internally disciplined enough to defend new trees from its own residents, passing nomads or other wanderers may do them in. To be successful, then, reforestation efforts often require a formidable administrative effort to protect the plants for years—not to mention the monitoring of timber harvesting and replanting activities once the trees reach maturity.

Village politics can undermine a program as well. An incident from Ethiopia a few years back presents an extreme case, but its lessons are plain. A rural reforestation program was initiated as a public works scheme to help control erosion and supply local wood needs. The planting jobs were given to the local poor, mostly landless laborers who badly needed the low wages they could earn in the planting program. Seedlings were distributed, planting commenced, and all seemed to be going well—until the overseers journeyed out to check the progress. They found that in many areas the seedlings had been planted upside down! The laborers, of course, well knew the difference between roots and branches; they also knew that given the feudal land-tenure system in which they were living, most of the benefits of the planting would flow one way or another into the hands of their lords. They were not anxious to work efficiently for substandard wages on a project that brought them few personal returns (Thomas, 1974).

In country after country, the same lesson has been learned: tree-planting programs are most successful when a majority of the local community is deeply involved in planning and implementation, and clearly perceives its self-interest in success. Central or state governments can provide stimulus, technical advice, and financial assistance, but unless community members clearly understand why lands to which they have traditionally had free access for grazing and wood-gathering are being demarcated into a plantation, they are apt to view the project with suspicion or even hostility. With wider community participation, on the other hand, the control of grazing patterns can be built into the program from the beginning, and a motivated community will protect its own project and provide labor at little or no cost.

An approach like this—working through village councils, with locally-mobilized labor doing the planting and protection work—is now being tried in India. This approach too has its pitfalls; Indian villages are notoriously faction-ridden, and the ideal of the whole community working together for its own long-term benefit may be somewhat utopian. But if it can get under way on a large scale, the national program in India may succeed. Once given a chance, fast-growing trees bring visible benefits quickly, and they just could catch on. The Chinese have long used the decentralized, community labor-mobilizing approach to reforestation, apparently with moderate success.

Alternative Fuels

Whatever the success of tree-planting projects, the wider substitution of other energy sources where wood is now being used would, if feasible, contribute greatly to a solution of the firewood problem. A shift from wood-burning stoves to those running on natural gas, coal, or electricity has indeed been the dominant global trend in the last century and a half. As recently as 1850, wood met 90% of the fuel needs of the United States, but today in the economically advanced countries, scarcely any but the intentionally rustic, and scattered poor in the mountains, chop wood by necessity anymore. In the poor countries, too, the proportion of wood users is falling gradually, especially in the cities, which are usually partly electrified, and where residents with much income at all may cook their food with bottled gas or kerosene. Someone extrapolating trends of the first seven decades of this century might well have expected the continued spread of kerosene and natural gas use at a fairly brisk pace in the cities and into rural areas, eventually rendering firewood nearly obsolete.

Events of the first half of the 1970s, of course, have abruptly altered energy-use trends and prospects everywhere. The most widely overlooked impact of the fivefold increase in oil prices, an impact drowned out by the economic distress caused for oil-importing countries, is the fact that what had been the most feasible substitute for firewood, kerosene, has now been pulled even farther out of reach of the world's poor than it already was. The hopes of foresters and ecologists for a rapid reduction of pressures on receding woodlands through a stepped-up shift to kerosene withered overnight in December 1973, when OPEC announced its new oil prices. In fact, the dwindling of world petroleum reserves and the depletion of woodlands reinforce each other; climbing firewood prices encourage more people to use petroleum-based products for fuel, while soaring oil prices make this shift less feasible, adding to the pressure on forests.

The interconnections of firewood scarcity, ecological stress, and the broader global energy picture set the stage for some interesting, if somewhat academic, questions about sensible disposition of world resources. In one sense it is true that the poor countries, and the world as a whole, have been done a favor by the OPEC countries, who through price hikes and supply restrictions are forcing conservation of a valuable and fast-disappearing resource, and are not letting the poor countries get dangerously hooked on an undependable energy supply. In a sensibly organized world, however, taking into account the total picture of energy, ecology, and food, the oil distribution picture would look far different. The long-term interest in preserving the productive capacity of the earth and in maximizing welfare for the greatest number of people might argue for lower prices and a rapid *increase*, not a halt, in the adoption of kerosene and natural gas in the homes of the poor over the next two decades. This in turn would be viable for a reasonable time period only if the waste and comparatively frivolous uses of energy in the industrial countries, which are depleting petroleum reserves so quickly, were cut sharply. It is not so far-fetched as it might first seem to say that today's driving habits in Los Angeles, and today's price and production-level decisions in the Persian Gulf, can influence how many tons of food are lost to floods in India, and how many acres of land the Sahara engulfs, in 1980.

Fossil fuels are not the only alternate energy source being contemplated, and over the long term many of those using fire-wood, like everyone else, will have to turn in other directions. Nothing, for example, would be better than a dirt-cheap device for cooking dinner in the evening with solar energy collected earlier in the day. But actually developing such a stove and introducing it to hundreds of millions of the world's most tradition-bound and penniless families is another story. While some solar cookers are already available, the cost of a family unit, at about $35 to $50, is prohibitive for many since, in the absence of suitable credit arrangements, the entire price must be available at once. Furthermore, no inexpensive means of storing heat for cloudy days and for evenings has yet been devised (Moumoumi, 1975; National Academy of Sciences, 1972; Daniels, 1974; Hayes, 1975).

Indian scientists have pioneered for decades with an ideal sounding device that breaks down manures and other organic wastes into methane gas for cooking and a rich compost for the farm. Over eight thousand of these bio-gas plants, as they are called, are now being used in India. Without a substantial reduction in cost, however, they will only slowly infiltrate the hundreds of thousands of rural villages where the fuel problem is growing. Additionally, as the plants are adopted, those too poor to own cattle could be left worse off than ever, denied traditional access to dung but unable to afford bio-gas (Prasad et al., 1974; Markhijani et al., 1975). Still, it is scientific progress with relatively simple, small-scale devices like solar cookers and bio-gas plants that will likely provide the fuel source of the future in most poor countries.

In terms of energy, Nepal is luckier than many countries in one respect. The steep slopes and surging rivers that cause so many environmental problems also make Nepal one of the few remaining countries with a large untapped hydroelectric potential. The latent power is huge, equaling the hydroelectric capacity of Canada, the United States, and Mexico combined. Exploitation of this resource will be expensive and slow, but would relieve some of the pressures being placed on forests by the larger towns of Nepal and northern India. On the other hand, cheap electricity will only partly reduce firewood demands, since the electrification of isolated villages in the rugged Himalayas may never be economically feasible.

Back to the Basics

Firewood scarcity will undoubtedly influence governmental perceptions of the population problem in the years ahead. It spotlights the urgency of slowing population growth through action on several fronts: making family planning services universally available; encouraging the liberation of women from traditional roles; meeting the basic social needs such as rudimentary health care, adequate nutrition, and literacy that are usually associated with reduced fertility; and reorienting social and economic incentives to promote smaller families.

The firewood crisis, like many other resource problems, is also forcing governments and analysts back to the basics of man's relationship with the land—back to concerns lost sight of in an age of macroeconomic models and technological optimism. Awareness is spreading that the simple energy needs of the world's poorest third are unlikely ever to be met by nuclear power plants, any more than their minimum food needs will be met by huge synthetic protein factories.

Firewood scarcity and its attendant ecological hazards have brought the attitude of people toward trees into sharp focus. In his essay "Buddhist Economics," E. F. Schumacher (1973) praises the practical as well as esoteric wisdom in the Buddha's teaching that his followers should plant and nurse a tree every few years. Unfortunately, this ethical heritage has been largely lost, even in the predominantly Buddhist societies of Southeast Asia. In fact, most societies today lack an ethic of environmental cooperation, an ethic not of conservation for its own sake, but of human survival amid ecological systems heading toward collapse.

This will have to change, and fast. The inexorable growth in the demand for firewood calls for tree-planting efforts on a scale more massive than most bureaucrats have ever even contemplated, much less planned for. The suicidal deforestation of Africa, Asia, and Latin America must somehow be slowed and reversed. Deteriorating ecological systems have a logic of their own; the damage often builds quietly and unseen for many years, until one day the system collapses with lethal vengeance. Ask anyone who lived in Oklahoma in 1934, or Chad in 1975.

References

Adeyoju, S. K., and E. N. Enabor, 1973. *A Survey of Drought Affected Areas of Northern Nigeria.* University of Ibadan, Department of Forestry, November, p. 48.

Bishop, B. C. and L. M., 1971. Karnali, Roadless World of Western Nepal. *National Geographic*, 140(5), November, p. 671.

Brown, L. H., 1973. *Conservation for Survival: Ethiopia's Choice.* Addis Ababa: Haile Selassie I University. Pp. 63-64.

Daniels, F., 1974. *Direct Use of the Sun's Energy.* New York: Ballantine Books. Originally published in 1964.

Delwaulle, J. C., 1973. Désertification de l'Afrique au sud du Sahara. *Bois et Forêts des Tropiques.* 149 (Mai-Juin): 14.

DuBois, V., 1974. *The Drought in West Africa.* American Universities Field Staff, West African Series. 15(1).

Government of India, 1973. *Interim Report of the National Commission on Agriculture and Social Forestry.* New Delhi: Ministry of Agriculture. August, p. 37.

Hayes, D. A., 1975. Solar power in the Middle East. *Science,* 188 (27 June 1975): 1261.

Le Houérou, H. N., 1970. North Africa: Past, Present, Future. In *Arid Lands in Transition,* ed. H. E. Dregne. Washington, D.C.: American Association for the Advancement of Science.

Makhijani, A., with A. Poole, 1975. *Energy and Agriculture in the Third World.* Cambridge, Mass.: Ballinger Publishing Co. (especially chapter 4).

Mason, J. A., 1957. *The Ancient Civilizations of Peru.* Harmondsworth: Penguin Books. P. 137.

Moumoumi, A., 1975. Potentials for Solar Energy in the Sahel. *Interaction* (Washington, D.C.) 3(10), July.

National Academy of Sciences (NAS), 1972. *Solar Energy in Developing Countries: Perspectives and Prospects.* Washington, D.C.: Office of the Foreign Secretary, March.

Openshaw, K., 1974. Wood fuels the developing world. *New Scientist,* 61(883), 31 January.

Prasad, C. R., K. Krishna Prasad, and A. K. N. Reddy, 1974. Bio-gas Plants: Prospects, Problems and Tasks. *Economic and Political Weekly* (New Delhi) 9(32-34). Special number, August.

República del Peru, 1974. *Lineamientos de política de conservación de los recursos naturales del Peru.* Lima: Oficina Nacional de Evaluación de Recursos Naturales and Organización de los Estados Americanos. May, p. 18.

Richardson, S. D., 1966. *Forestry in Communist China.* Baltimore, Md.: Johns Hopkins University Press. Pp. 14, 66.

Schumacher, E. F., 1973. Buddhist economics. In *Small is Beautiful: Economics as if People Mattered.* New York: Harper & Row.

Thomas, J. W., 1974. Employment creating public works programs: observations on political and social dimensions. In *Employment in Developing Nations,* ed. E. O. Edwards. New York: Columbia University Press. P. 307.

UN/FAO, 1967. *Wood: world trends and prospects.* Rome: FAO Basic Study No. 16.

UNDP/FAO, 1973. *Forestry research, administration, and train-
ing (Arsil, Iraq), Technical Report I,* "A Forest Improvement
Programme." UNDP/FAO IRA 18 TR 1, IRQ/68/518. Rome:
FAO.

UNDP/FAO, 1969. *Agricultural survey and demonstration in
selected watersheds, Republic of Korea,* Vol. 1, General Report. FAO/SF: 47/KOR 7. Rome: FAO, p. 17.

U.S. Department of Interior, 1964. *Land and water resources of the
Blue Nile Basin, Ethiopia. Appendix VI. Agriculture and
Economics.* Washington, D.C., p. 12.

Vermeer, D. E., 1970. Population pressure and crop rotational
changes among the Tiv of Nigeria. *Annals of the Association
of American Geographers.* 60(2) June, p. 311.

4

Ecological Deterioration: Local-Level Rule-Making and Enforcement Problems in Niger

James T. Thomson

Officials live off the people. The people live off the bush. Now the officials have prohibited us from living off the bush. What are we supposed to do? We just go into the bush and we steal. If the forester catches us, he catches us.

This villager's comment neatly summarizes the dilemma of ecological deterioration plaguing his homeland in the Sahelian region of south central Niger, West Africa. Rural people, prodded by daily necessity, violate rules created to protect overtaxed environmental resources: they simply see no crime in trying to keep body and soul together today, whatever the regulations say. At the same time, their efforts constantly deplete the already diminished ground cover—grasses, bushes and woodstock—of the Sahel, baring it to the onslaught of severe weather conditions and the creation of man-made local deserts from the Saharan fringe far south through the Sahel and even into the savannah zone.

This outcome is by no means inevitable. But to avoid it, incentives now leading people to ignore the threat of long-term disaster in favor of short-term survival must be altered. In achieving this outcome, national conservation policies are of precious little relevance unless they induce changes at the local level where daily, pressing needs override rules ostensibly designed to conserve wood resources. By itself, stringent enforcement of these rules, perceived to be unjust by villagers, is an inadequate solution (Raeder-Roitzsch, 1974).[1] Enforcement efforts undertaken in isolation, without concern for expanding the overall wood supply, merely provoke villagers to further evasion or to bribing officials. The result is unregulated exploitation of the ground cover which, while a replenishable resource, cannot survive in the face of escalating demand unless effective supplies are increased.

The policy problem thus is to facilitate a productive relationship with the environment. To clarify how this goal might be approached, we will first analyze the economic and legal factors underlying the current ecological tragedy, then examine the local failure of national policies to remedy the problem exclusively by regulating consumption of ground cover, and finally offer two tentative reform proposals that could marginally improve the situation. The analysis stresses the local side of the problem and the necessity for creating a local capacity to devise and implement solutions, including acceptable rules and effective rule enforcement procedures. This emphasis is not intended to deny or obscure the wider regional, national, and international aspects of environmental deterioration in the Sahel. But it does seek to underline the critical, and often neglected, local component of the problem.

Economic and Legal Aspects of Ground Cover Use

The problem of environmental deterioration in Inuwa District[2] hinges on the economic and legal relationships that currently govern the use of wood in the jurisdiction. The wood stock and other components of the ground cover are effectively nationalized common property goods. They are subject to control by officials representing the Nigerien national community, rather than being owned individually by particular villagers. In consequence, users take what they need when they need it and can get away with it, without regard to the effects their actions may have on maintenance of the wood stock as a resource held in common by members of the national community. As long as ground cover is adequate this behavior poses no problem—this was the case during the precolonial and most of the colonial eras, when population was too sparse to seriously tax the available ground cover. However, when effective demands on the resource outstripped the supply, individual demands became competitive. Now, as individuals appropriate or destroy more and more ground cover, others find it increasingly expensive or even impossible to meet their own needs from the same resource.

Because the wood stock is viewed as a common property, villagers consider its preservation an official problem. Worse yet, an individual villager's efforts to preserve or increase ground cover would be senseless. He might plant trees, or refrain from destroying standing timber, but he gains thereby no assurance that he will harvest the fruits of his efforts or abstinence. Another user may well appropriate the wood before he decides to use it, leaving him with no return on his outlay. Thus, everybody uses, nobody replaces, and use becomes abuse. Demands on the wood stock escalate to the point where it may well be destroyed in the not-too-distant future.

Use Patterns

These are pessimistic assertions. To justify them we must briefly identify the pattern of uses to which the wood stock in Inuwa District is subject. Local variations do exist throughout the West African Sahel, but the general pattern holds, as does the range of associated problems (Raeder-Roitzsch and Zenny, 1974). Major factors determining this pattern are climatic conditions and the needs of the partially complementary, partially competitive farming and herding ecologies dominant in the Sahelian zone.

Consumptive Uses: Farming

In Inuwa District, village-dwelling agriculturalists[3] derive the staples of their existence from rainy season farming, though they keep livestock and pursue dry season trades as well (Nicolas, 1960, 1962). Traditionally, Inuwa farmers practiced a type of rotating fallow agriculture which left a substantial portion of the land with some form of ground cover. But expanding population and extensive agricultural techniques have combined to bring under cultivation almost all arable land in the district. Thus farming has taken on the characteristics of a destructive use of ground cover. Now, many families cannot meet the fallowing requirements of the traditional system on their limited acreage, and soils are deteriorating (République du Niger, 1971).

The spread of field agriculture has drastically reduced the local wood stock: in clearing ground for cultivation farmers used to cut all but the largest trees, then often fired the bush, killing most tree seedlings. While farmers no longer fell mature trees before planting, seedlings are still destroyed unless special precautions, involving annoying extra effort, are taken to preserve them.

In addition to land clearance, the sedentary population consumes large amounts of wood for fuel and construction. The steady demand for wood and wood products used in cooking, heating, housing and storage construction, tools, and fencing increasingly depletes the wood stock. Enclosures pose a special problem. Thorn branches are the only material available to erect stock-proof fences on a large scale. Even then it is possible to enclose large areas only when a number of farmers can cooperatively fence a common terrain (Nicolas, 1962).[4] Land tenure patterns of two of the three major sedentary groups render collective fencing infeasible. Members of these groups thus find it preferable to employ village herders during the growing season to confine sedentary-owned livestock in grazing commons. During the long dry season, however, village fields revert to common pasture land, and livestock wander freely in search of fodder. Those who garden during the dry season must barricade their plots with thorn fences to exclude hungry animals.

Consumptive Uses: Herding

Through much of the year farmers and herders coexist uneasily within Inuwa District. Extension of field agriculture into areas traditionally considered rangeland has exacerbated the tension, as herders see their own resource base correspondingly reduced. During the June-September wet season herders[5] avoid conflicts by taking their livestock north into the Saharan fringelands. When the rains end, they descend again to the southern regions. If they arrive before farmers have put up the summer's harvest and allow their animals to devour unstored field crops, vicious fights can erupt. These the farmers generally lose. On the other hand, once crops are safely stored, farmers want livestock to graze their fields: manure is the major source of fertilizer for lands not currently fallowed (Raulin, 1960).[6]

By February or March, at the height of the dry season, the last edible stalks have been consumed. From April to June, herds must survive on a miserably thin diet gleaned from barren fields. To keep them alive, herders lop entire branches from trees, which their cattle then strip of foliage. In the process, however, trees may be reduced to lifeless trunks. Occasionally herders also fire bushland in the dry season, hoping to trigger an early growth of grass to sustain their animals until the rains. This practice destroys tree seedlings. They are also exposed, despite their thorns, to intense grazing pressure, not only by the nomads' cattle and sheep, but also by villagers' goat herds.

As the population continues to expand, the combined impact of herding and farming ecologies overwhelms the fragile Inuwa environment. Denuded soils are exposed to wind and water erosion. Stripping wood cover from the land not only stops the natural formation of humus but reduces the value of manure by exposing it to the sun for long periods before the rains decompose it and wash it into the soil.

Of potential assistance in stemming this trend is one species of large thorny acacia tree, *Faidherbia albida,* because of its anomalous growing cycle. It bears its foliage during the dry season, when other trees stand leafless, and drops them during the rains. As a shade source in the dry season it attracts animals; during the wet season, it offers farmers a strategically fertilized, sunny spot when other trees, then in foliage, shade out crops (Weber, 1971).

However, the inability to effectively exclude animals and humans from any but garden-sized plots during the dry season means that no one has any incentive to improve the quality of the wood stock on his own fields. The acacia suffers particularly in this respect since its dry season foliage makes it a prime target for fodder-chopping herders.

This combination of legal regulations and competitive economies has convinced most residents of Inuwa District that conservation efforts, while they may be nice in the long run, are pointless in the short run. Daily experience confirms that ground cover—grasses, brush, and trees—is an unregulated common property subject to exploitation by any and all. Defending that environment, or managing it to the mutual benefit of all, is out of the question.

Theoretically, this should not have been the case. To understand why, it is necessary to review French colonial attempts to conserve wood resources in the West African colonies. French colonial officials recognized, as early as the mid-1930s, the potential gap between supply of ground cover and wood products and demand for them. A conservation program begun then was extended in 1949. Decrees authorized administrative control of grazing, burning, and woodcutting practices in an effort to maintain a reserve of woodland (Hailey, 1957).

Logic of Regulated Common Property Resource

The thrust of this French conservation effort centered on a negative goal, the prevention of overgrazing and woodcutting; in many cases, any exploitation of ground-cover resources was absolutely prohibited. Forest management for maximum sustained yield was a goal apparently not considered (Raeder-Roitzsch and Zenny, 1974). The French approach in its simplest form—declaring a forested area off limits to all users—nationalized ground cover. Villagers, through the officials who governed them, were granted certain rights, consisting essentially of the reciprocal duty of all others to avoid exploiting the wood stock.[7] Likewise, each individual was himself "under a duty" to avoid infringing on the wood stock. Thus a hands-off policy prevailed all around, guaranteed and enforced by a colonial forestry service. The individual's right amounted to a guarantee that while he could make no use of the common property resource, nobody else was at liberty to do so either. Violators were heavily fined when apprehended (Territoire du Niger, 1957).[8]

Frequently, this absolute prohibition did not apply, and limited and controlled exploitation of the wood stock was permitted through a licensing system. The license purchased from the forestry officials authorized the holder to exploit the stock within the limits specified by the permit (*Le Niger*, 1973), and exposed all other potential users of the common property to the consequences of that usage.[9] Theoretically these risks were to be limited by forestry service officials to an acceptable level. That is, they would insure that the wood stock was never exploited beyond the point where it could regenerate itself by natural processes.

Insofar as this policy helped to prevent the total destruction of ground cover and the creation of artificial deserts in the Sahel, while maintaining local sources of wood as well, it was a valuable program even for those unwilling or unable to purchase cutting permits. But the regulated lands included many areas villagers considered theirs, and they understandably felt the entire program to be unjust in its conception. Those now forced to buy licenses before cutting, or fined when they felled trees without permission, resented the new regulations, particularly since they were usually imposed unilaterally by the colonials without prior consultation to determine local needs or customary rights.

Nonetheless the French persisted, evidently with some short-term success in dampening demand. They organized a forest service in each of the French West African colonies, primarily to enforce regulations. Colonial forestry executives controlled a bureaucracy that eventually extended down to the county (*cercle*) level. Field agents worked out of these administrative centers and, assisted by quasi-official retainers of district chiefs, policed their jurisdictions. They determined violations, assessed fines, and generally tried to prevent inroads on the forest preserves.

The same conservation system was continued intact by the independent regimes which acceded to power in 1960 in French West Africa. But enforcement proceedings have slackened markedly; the regulated common property now effectively reverts to the status of an unregulated common property.

Logic of Unregulated Common Property Resource

Under the permit system, the right of each user of the common-property wood stock consists, as indicated, in the prohibition of overcutting by other users. Each is under exactly the same duty not to fell timber without a permit. When the permit system is loosely enforced or simply disregarded, individual rights, duties, liberties, and exposures concerning the wood stock are substantially modified. Rights are obviously eroded in value. When the duty to avoid cutting can be ignored by some or all, without regard to maintaining a balance between supply and demand, the individual who continues to observe his "duty" merely imposes a short-run economic loss on himself without any guarantee that the wood stock will be preserved indefinitely as a source of wood products. Duties neither enforced nor observed are meaningless. Instead, liberties to exploit become the critical factor. Effective policy, as implemented by officials, then either encourages uncontrolled exploitation of the wood stock, or does little to restrain it. This in turn unleashes the negative dynamic of a tragedy of the commons (Hardin, 1968) in which the relationships among all parties become mutually destructive. Deterioration of ground cover and anthropogenic desertification in the Sahel will benefit neither of the two major groups of users, the farmers nor the herders. Since the urban economies of most of the Sahelian states are closely tied to the productivity of their rural economies, city dwellers and national governments can expect to suffer in the end from a policy which permits desertification.

How does one account for breakdowns in enforcement, in light of such grim predictions? Several factors, already alluded to, must be considered. First, rural dwellers clearly have to survive today before they worry about tomorrow. Second, within the existing system of legal regulations, local people have no strong incentives to preserve the wood stock, much less expend effort in order to increase it. Under the same system, however, county and district forestry officials can benefit by authorizing systematic deterioration of the wood supply. Several cases of confrontations between rural dwellers and forestry officials will indicate the logic that underlies their strategy.

National Policy in Practice: Local-Level Forestry Cases

The following cases record typical disputes with forestry officials experienced by villagers in Dajin Kowa, a community located in Inuwa District. To enforce conservation regulations the forestry agent makes bimonthly tours through Inuwa District. Generally he rides with a traditional policeman assigned to him as a permanent assistant by the Inuwa District chief. The policeman detects infractions and in principle identifies delinquents for the agent. The latter then assesses the damage and writes a ticket specifying the fine. If official procedures are followed, he forwards the ticket to the Mirria County conservation office. From there it is returned via the Mirria subprefect—chief administrator for the jurisdiction and a national government cadre—to the Inuwa district chief. A policeman delivers it to the offender, who finally pays the fine at the Mirria office.

Official procedures are often ignored, however, as the following cases indicate. The first occurred in 1968.

A father and son were trimming acacia branches to repair the thorn fence surrounding their garden. Two horsemen rode up: the agent and a district policeman. They announced the forestry code prohibited trimming acacias and issued a ticket.

Later the son took 250F CFA and followed them to a neighboring village. There he tried to bribe the agent.[10] The latter refused, saying that 2,500F CFA would be more appropriate in view of the damage.

Back in Dajin Kowa the gardener collected 1,500F CFA. He then went to the district seat to find the Mirria County councilor from Inuwa District. Having explained his difficulty and having secured a promise of assistance, he went home.

At daybreak the councilor and the gardener set out for the community in which the agent had said he would spend the day collecting fines. Once there, the two enlisted the support of a local school teacher, a friend of the gardener. Together they sought out the agent. The councilor offered the 1,500F CFA, but the agent rebuffed him, saying the gardener knew the amount was 2,500F CFA. The teacher and the councilor were not dissuaded, however; and at sundown, after a day's pleading, the agent took the 1,500F CFA and dropped the matter.

According to the gardener, in such situations the violator must get the backing of a local notable to serve as intercessor.[11] If that person has influence with the agent, because of official position, family ties, etc., and is persistent, a favorable arrangement can be achieved. District councilors demand nothing for such services, he said: they should aid villagers in trouble, and usually do. Sule, the Dajin Kowa village political committee president, confirmed the assertion, indicating that with party help any sort of difficulty could be settled[12] (Higgott and Fugelstad, 1975).

President Sule was personally involved as a violator in a case which attests the accuracy of his statement.

In late 1970 Sule built a new silo, constructing the base from an acacia tree he cut in his own field. About the same time, Sule said, a nomad cut another acacia in his field, in order to fodder his cattle.

Shortly thereafter, as Sule and his wife were winnowing peanuts in the field, the forester and his police assistant appeared. Sule admitted his violation. The agent issued a ticket for *both* trees and the president took it without protest. He was told to meet the agent in another village, or else the ticket would go to Mirria and Sule would pay the full amount, about 4,000 to 6,000F CFA, Sule thought.

The president left at once to contact the other Inuwa District county councilor. He found him collecting taxes with the district chief and asked for his help as intercessor. The councilor refused, being preoccupied with tax matters, but offered to send in his stead a district policeman. The one he appointed was a Dajin Kowa resident whose father was friendly with the forester and put him up when he stayed in the village. As such, the policeman would be a very suitable intercessor. Sule handed him his 500F CFA bribe for the agent and the two set out for the appointed meeting.

At the outskirts of the village Sule stopped. The policeman went on to deliver the councilor's greetings and Sule's 500F CFA; both were accepted, but with the warning that Sule had best avoid acacia trees in the future.

About a year later President Sule saw the same herder in his field again, chopping limbs from another acacia. To avoid further trouble, he told the agent. The latter said the herder in question had often been fined, but to no avail: he continued cutting as he pleased.

These two cases suggest how the regulation system really works, and we will consider the implications shortly. For the moment, a review of the administrative history of the Inuwa District forest conservation program is in order, to illustrate more fully the code enforcement powers of the forester's assistant, and the extent to which he can abuse them.

At the end of the colonial era the Inuwa District chief appointed a young relative, Illo, as assistant. He profitted handsomely from the office during the year he held it, by approaching ticketed individuals after the agent left the district with offers to fix their fines for a bribe. According to fellow officials, Illo occasionally even extracted sheep, animals of substantial value, from violators.

But he neglected to fix the fines. Finally, a group of villagers complained to the agent and the district chief. All had lost money through Illo's shake-downs. The chief refunded 15,000F CFA to them and, fearful that his relative's behavior would involve him in difficulties as well, removed Illo from the post. The latter remains a member of his retinue, however, and still works for the chief.

Of the next assistant, his fellow district officials reported scornfully, *la la la gareshi* (timid soul): he feared the risks of fixing fines and so didn't. Four years he held the post, until age forced his retirement. The job then went to another district policeman, Maigemu, who held it seven years. In a 1971 interview Maigemu stressed that he regularly fixed fines for violators. He viewed code enforcement as a problem of political economy and official authority:

> A herder caught trimming an acacia pays 6,000F CFA. Guilty farmers are likewise fined, but generally much less than herders. If they admit their guilt and plead for relief, the forester and his assistant destroy the ticket in return for a 1,000 to 1,500F CFA payment. Nonbribers pay in full.
>
> According to Maigemu, farmers can trim low branches of acacia to facilitate cultivation of underlying areas, but trees may be felled only with permission. Those who cut without asking are actually challenging officials' authority and must be rebuked.
>
> Nonresidents caught cutting without permits face severer treatment than locals, and rich violators are assessed more than poor ones, he asserted. Often the poor are merely lectured, forced to surrender their tools, and released. Such leniency was justified, according to Maigemu, because it was pointless to persecute neighbors.

Maigemu's statements were tested in a case that occurred shortly after he made them, and as a result of which he was fired.

> The forester accused his assistant, in the district chief's court, of having embezzled 10,000F CFA over the past two years, despite the fact that the agent paid him roughly that amount annually in salary and reimbursements for Maigemu's personal taxes. The inquiry revealed that Maigemu had embezzled not only the fines, but the tax monies—the chief had in fact paid all his taxes over the past seven years—and additional funds as well. In the end, the charges were dropped: Maigemu lost only his post and the horse which the forester had loaned him. He also continues as a member of the chief's entourage.

After the trial other district officials said Maigemu's favorite strategy was to discover code violators and then, if the agent failed to detect the damage on his next tour, to demand bribes from guilty parties as the price for not revealing their offenses. Another informant, a part-time woodcutter in Inuwa District, noted that those intending to fell trees often sought out Maigemu before entering the bush and paid what he demanded for protection. Bribes ranged from 50F CFA up, depending upon amount and type of wood involved. Refusal to pay could lead to greater expenses when the forester arrived, according to this informant.

Analysis of Cases

Two preliminary comments are in order: first, the cases reflect the poverty and lack of available alternatives already discussed, which force local people to raid the Dajin Kowa wood stock. Second, villagers have no say in setting rules governing wood usage: outside officials dominate the process of law making.

The cases suggest several tentative conclusions, though they provide no conclusive evidence for them. First, the Dajin Kowa village regime plays no more part in implementing the forestry code than it does in drafting it. Village officials have no authority to catch violators, nor do they seem to have any desire to do so. Enforcement responsibility lies entirely with authorities from outside the village. Second, non-village officials are usually able and willing to manipulate formal regulations to their own benefit by shaking down villagers arrested for violations of the forestry code. Third, villagers willingly bribe officials if they can, to reduce their own immediate costs when apprehended, without regard to the long-term implications of their actions. Fourth, under the circumstances, villagers accurately view wood production and consumption as a zero-sum game. The wood stock is effectively a deregulated common property resource and, in the struggle to meet their daily needs, villagers rationally reject any responsibility for conservation.

Let us examine these points more fully. First, the village regime does not count when it comes to enforcing the forestry code. Forest conservation is legally defined as a national responsibility; villagers have developed no effective organization for coping with the problem. Moreover, code enforcement proceedings frustrate village initiative. The forestry agent assigned to patrol Inuwa District and his assistant preempt the entire process. The agent depends upon "his" policeman for local knowledge and enforcement during his absence. Both agent and assistant are informally constrained to justify their administrative offices by fining some violators, but there is no (economically feasible) administrative means of determining if they fine them all.

Second, both agent and assistant exercise substantial enforcement powers. They can compel exploiters of the common property to abide by formal—or informal—regulations governing its use. Within very wide limits, the agent is immune from control by all but his own superior at the county level. Cases turned up no attempt by either commoners or Inuwa District authorities to have him sanctioned for failure to enforce code provisions. Rather, both commoners and Inuwa officials induce him to manipulate regulations.

The police assistant operates within ostensibly narrower limits of immunity. Violators once succeeded in having an aide sanctioned for abusing his authority by promising to fix fines. But the impetus for challenging him lay in his failure to "arrange things," not in his willingness to trade bribes for code manipulations. The forester independently dismissed another policeman—Maigemu—for excessive bribe-taking as well. Yet the Inuwa District chief kept both men in his service and they continue to enjoy the standard benefits involved.[13] A prudent assistant can thus exploit his position for personal profit over long periods. And he enjoys extensive *de facto* immunity from control because his superior makes only brief, infrequent tours through the district. During his long absences the assistant takes command and structures enforcement in light of his own immunity from control, and opportunities to profit, by selectively exercising his enforcement powers. He sells a valuable service—liberty to exploit the common property—instead of enforcing the duties users are theoretically required to observe concerning purchase of harvesting permits by levying fines for violations.

These illegal transactions occur within the district regime. They establish the villager's *ex ante* or *ex post facto* liberty to exploit the common property and the exposure of all other users to any externalities his use may generate. Evidently when the assistant neglects to enforce a duty against one user he decreases the right—devalues it, in economic terms—of all others who use the common property and increases their exposure. At the same time he impedes the maintenance of the environment for the public good by authoritatively permitting a user to mine the existing wood stock without generating revenues to fund additional regulation or reforestation activities. Finally, of course, he swells his own income.

Third, the cases reveal that for commoners the forestry code remains at best a nuisance and at worst a generator of serious financial burdens. They seek whenever possible to evade provisions of the code. When they cannot entirely evade them, commoners seek to avoid full application. They either bribe beforehand for protection or, if charged with infractions of the code, try to have their fines reduced by the sanctioning official, approaching him through another official who has status and/or enjoys the forester's respect. Individuals who lack connections with Inuwa officials, nomads, for example, either pay a higher price for *ex post* liberties of exploitation or find themselves subject to the full letter of the national code.

Except when they fear they may be held responsible for violations committed by others, commoners make no attempt to have officials enforce code provisions against other illicit users of the common property. They fear not only the ill will such actions would evoke among fellow villagers, but the animosity they risk incurring on the part of enforcement or other officials who might later be in a position to retaliate—if, for example, the complainant himself is caught violating the code, needs more time to collect his tax payments, or becomes involved in a lawsuit tried in the district court.

The bribe bargain thus directly discourages commoners from claiming their rights. Officials, who might already have struck a deal with the supposed violator, would resent a villager's openly concerning himself with their affairs. They consider that authoritative transactions involving the common property concern only the officials and persons who are direct parties to them. Thus woodcutting is treated as though it were a private matter between officials selling code manipulations and villagers buying locally authoritative liberties. They are not viewed as public transactions involving multiple interests in a common property resource.

Therefore, barring institution of much more effective regulation procedures coupled with supply increases, a tragedy of the commons is definitely indicated. The value of wood will rise with further depletion of the resource, leading to both increases in the bribe price charged for the liberty of exploiting it and a more intensive clandestine scramble to appropriate the remaining stock.

The incentive for any single user to invest time and effort in preserving or reconstituting the stock is correspondingly minimized: the fruits of his labor would be converted into the booty of others' forays into the bush to "steal" wood.

Under these conditions existing regulatory practices will neither dampen basic fuel and construction demands nor increase supply. Only as the total environment erodes, forcing many Inuwa District residents to emigrate, will demand fall off. At that point, however, the productive capacity of the district will have been substantially, perhaps irreparably, impaired. This suggests the degree of exposure under which each user of the common property labors with respect to the liberty of all other users in the exploitation of that property.

Reform Proposals

Analysis of the cases suggests that the fundamental problem underlying deterioration of the Sahelian environment is the difficulty of regulating demand on the wood supply. Until this difficulty can be dealt with, efforts to conserve the supply, let alone increase it to a level consonant with effective demand, will continue to founder under a quasi-universal system of law breaking, as has the national regulatory program. This system of code violation and manipulation, which vitiates the impact of a regulatory program almost entirely, is supported by villagers because they know they cannot meet their need for wood products by acting within the bounds of the law.

If the problem in maintaining the wood stock as a common property resource lies in a fundamental lack of enforcement, then remedies must make regular enforcement feasible. To achieve this, the perceptions, and thus the motivations, of villagers must be changed. To what extent is it possible to do so by manipulating rules governing use of the wood stock?

Presumably any remedy must modify villagers' views of the problem by addressing itself to their short-term as well as long-term needs. Initially, villagers must be convinced that their immediate wants can still be satisfied; this implies a reliable increase in supply, so wood will be available to meet indispensable needs. Simultaneously, enforcement procedures must be strengthened: given continuing competition among multiple users and uses, it can be assumed that rules seeking to modify use patterns *will be violated*, just as current code regulations are. Thus, the success of alternative approaches turns finally on the problem of *rule enforcement*. Without effective enforcement, an increase in supply is impossible. It follows that individualistic and collectivistic attempts to increase supply must be assessed in light of the probable locus and character of enforcement powers. We consider two such approaches, both of which assume that primary responsibility for enforcement is vested at the village level.[14]

Individual Ownership of Trees

First, control of the wood supply could be shifted to individual landowners, each gaining title to the trees on his property. Each could then supply his own needs or purchase wood from others.[15] All villagers, however, would be exposed to the liberty of any other villager to denude his own land of trees, endangering both wood supply and maintenance of a stable environment. This outcome would be avoided only if enough landowners found it advantageous to maintain a wood stock for personal use and/or sale as a cash crop.

Assuming an undiminished demand for wood, an individualistic supply system is probably workable if property rights in the wood stock are reasonably certain. This will be the case if individuals find enforcement proceedings readily available, inexpensive, and effective.[16] Will they? Relevant factors can be briefly canvassed.

Compared to the forester under the present conservation program, or even to his assistant, villagers have a distinct information advantage. Woodcutting activities on village lands rarely go unobserved, just as animal damage is rapidly discovered. If wood preservation is perceived by villagers as *their* problem, involving *their* property, instead of as a national government responsibility, new behavior is likely to evolve. By specifically identifying the property interests of villagers in particular trees, this system creates potentially powerful incentives—based on self-interest—for individuals to defend the wood stock. Many poachers, facing a high risk of detection, might be dissuaded from illegal cutting activities by the knowledge that their fellow residents would be inclined to act in defense of their woodstock, just as they defend their crops from marauding animals, or their land from would-be encroachers. This would clearly represent a marked change from the current conspiracy of silence that surrounds illegal wood cutting within Inuwa District villages.

Disputes would inevitably arise, however. If primary enforcement responsibility were located at the village level, would such claims be fairly settled by village officials? The issue is problematic. Where amounts at stake are small, some sort of settlement involving compensation or restitution might be arranged locally.[17] Furthermore, most local officials would probably find flagrant manipulation risky over the long run, although individual settlements might be manipulated by corrupt ones.

Perhaps a more important question is whether such conflicts could be settled locally at all. Factionalism at the village level and a largely effective national policy prohibiting use of self-help by villagers or physical force by village officials seeking to uphold regulations has to date vitiated the authority of many village leaders to the point where it is impossible for them to resolve disputes when good faith negotiation fails. Devolution of primary enforcement responsibility in forestry matters to the local level would have to include authorization to impose sanctions, subject to appeal, if local regimes are to be able to enforce regulations.[18] Otherwise, the only means to settle locally unresolved disputes would be recourse to an overriding regime.

Most cases settled on appeal would be resolved in the district chief's court, with limits to his enforcement power being established by eventual appeals to the Mirria County subprefect or the Zinder justice of the peace. Probably many claims would be foregone as too expensive to prosecute. Given district officials' predilection for the shakedown strategy, district court rulings in wood-cutting cases are likely to be both expensive and unpredictable. Costs of having defendants convoked to the district seat by traditional policemen, "greeting money" for the district chief acting as judge, possible bribes, etc., would dissuade most litigants from defending their wood stock against petty poaching. Substantial uncertainty about rights, duties, liberties, and exposures of woodcutters and wood owners would result. Permitted to continue unabated as they are now, such petty thefts in the end would discourage wood conservation practices, and might lead to large-scale destruction of the wood stock. Thus the relevance of readily available, inexpensive, and effective enforcement at the local level is apparent: without it, rule maintenance in most conservation conflicts is simply too costly to be effective in curbing destructive behavior.

Village-Level Ownership of Trees

An alternative reform measure might be to legally maintain the wood stock as a common property while shifting the locus of conservation efforts from the Mirria County and Inuwa District regimes to village authorities who would then be empowered to take and enforce collective decisions concerning the resource. Wood supply would be assured either by controlled cutting in the existing wood stock, by some form of joint production unit established on village land, or by a combination of both, depending on local conditions and predilections. Production under either arrangement might be enhanced by efforts to protect and manage natural regrowth, as well as by specific reforestation projects.

Once again, success of the reform would turn on the ability of villagers to develop effective enforcement procedures; otherwise, time and effort costs of disputing responsibility for protection and allocation of the supply would be prohibitive. Economies might be realized by employing a guard to police the supply and organize a nursery, special planting sessions, protection of nursery transplants and wild seedlings, and so forth.

Potential difficulties can be foreseen, including guards shaking down villagers caught cutting wood illegally, and villagers bribing the guard to obtain more than their just share of the wood supply. If such a situation became widespread, destructive competition among villagers for the remaining wood supply would intensify, with potentially catastrophic effects on the wood stock. However, villagers' proximity to the problem, their high level of information, and the fact that law abiders would immediately bear the ill effects of others' violations might promote a much more active insistence, locally, on rule enforcement. Appeals of such complaints to the district court, however, would probably face the difficulties detailed above in considering the individualistic alternative.

While local cliques might develop and successfully exploit the wood supply to the detriment of their fellow villagers, at least the effect of such developments would be confined to single communities, not spread over all villages in a district, as it is now when a corrupt forester and policeman control enforcement throughout the larger jurisdiction.

Finally, an advantage of either the individualistic or collectivistic alternatives outlined above would be to free Conservation Service personnel from their often corrupt, largely ineffective and thus expensive enforcement duties to concentrate on improving quality of seedlings, distributing nursery stock, and advising villagers on ways to increase yield while simultaneously maximizing the conservation contribution of the wood stock, such as wind breaks, game habitat, and soil fertility.

Summary

Contemporary environmental deterioration in the Sahel is largely the consequence of a national policy which in practice not only divorces local users from responsibility for conservation of the wood stock upon which their daily existence depends, but also creates a set of negative incentives that dissuade them from organizing to preserve it and thus serving their own long-term interests.

Several forestry code cases reveal how nationally-mandated attempts to control the wood stock and to dampen largely inelastic demand while failing to increase supply are aborted at the local level. A combination of factors permits and encourages national forestry agents to ignore code violations in return for bribes collected from villagers apprehended while illegally cutting wood, and likewise encourages villagers to reject all responsibility for conservation of the wood stock as they scramble to meet daily wood requirements by illegal means.

Solutions, it is argued, must achieve two goals: effective enforcement and an increase in supply, the former being a necessary condition for the latter. Two alternative reforms, individual ownership and village-level collective wood production, are assessed. Neither offers a foolproof remedy. Both, however, appear to facilitate better regulation and increased supply, essentially because they shift responsibility for enforcement to the local level, increasing villagers' identification with the local wood stock as their property and thus improving enforcement measures.

Acknowledgment

I thank Vincent Ostrom and Ronald Oakerson for their consistently insightful comments on earlier drafts of this article. I am also indebted to the Foreign Area Fellowship Program for funding field research in Niger (November 1970–March 1972) upon which this paper is based; to the Government of Niger for permission to undertake the study; and to many Nigeriens, including the personnel of the Centre Nigérien de Recherches en Sciences Humaines, Niamey, administrators, and the residents of Inuwa District, Mirria County, who were all both patient and helpful.

Notes

1. J. E. Raeder-Roitzsch, a professional forestry consultant, has reached similar conclusions about the forestry problem in Upper Volta.
2. "Inuwa" is a pseudonym for a rural district in Mirria County, Zinder State, Niger, where my field research was centered. Personal and place names have been altered to protect the identity of informants and others.
3. The principal agricultural groups in Inuwa District are Hausa, Barebari, Bugaaje (descendants of Twareg slaves), and some sedentarized Fulbe nomads.
4. The highly-organized tenure patterns of Bugaaje farms facilitate collective fencing; Hausa and Barebari fields, by contrast, are widely scattered, making enclosure with local materials virtually impossible.
5. Nomadic Fulbe are the main group in question here, although some Bugaaje also lead a seminomadic existence with their herds.
6. Inuwa people no longer use the laborious green-manuring technique prevalent in Niger's Maradi region, which permits continuous cultivation of fields without the necessity of fallowing.
7. The terminology adopted here concerning the legal concepts of *rights, duties, liberties,* and *exposures* is drawn from Commons.
8. By 1957 a conservation program had been implemented in Mirria County, but officials faced local resistance and slackened attempts to apply the regulations.
9. Applicants purchase cutting permits, renewable monthly for 700F CFA for 20 donkey loads of firewood. Permits to cut protected species range from 250 to 600F CFA per tree. Violators are officially liable to heavy fines, running to several thousand francs CFA. $1.00 = 250F CFA. Annual per capita income in Niger is currently about $60 to $80.
10. Villagers customarily use *gaisuwa,* presents in cash or kind, to initiate dealings with traditional officials. These are limited in value; legitimate *gaisuwa* for the Inuwa District chief do not

exceed 100F CFA. When asked, the gardener said he considered the 250F CFA he offered the agent a bribe, *el gaisuwa*, "daughter of *gaisuwa*."

11. The Hausa term for this role, *gozoma*—"midwife"—suggests the traumatic character, for villagers, of contacts with officials.

12. During the period when these cases occurred, Niger was governed by a single-party regime. The Parti Progressiste Nigérien-Rassemblement Démocratique Africain had established branches in most villages. In 1974 a military government took power in a coup, but little has changed in Nigerien politics since then.

13. On the evidence of other cases not here discussed, it would appear that district officials are dismissed by their chief only if they refuse to work for him. Embezzling funds destined for the chief, at least within broad limits, to say nothing of shake-down operations, are thus permissible actions.

14. One might instead assume primary enforcement responsibility to be formally switched from the county to the district level. The foregoing cases and analysis based on them suggest, however, that this would not materially modify the lax enforcement procedures which characterize the existing forestry program, and thus the assumption appears less interesting than that of local control. Whether local control is politically feasible in the current Nigerien political context, or in other states of the Sahelian belt, appears doubtful to me in light of the past performance of both civilian and military regimes. However, I explore the local control assumption as the most interesting one from the policy perspective of preventing environmental deterioration. At the least the ensuing discussion illustrates a set of potentially effective remedies for the current dilemma.

15. Equity problems would arise with this approach; not many district residents are entirely landless, but some are land poor and lack fallow fields whose wood resources could be exploited. At least during an initial period their needs would have to be met through redistribution if the new arrangement were to gain legitimacy.

16. Demsetz (1967) suggests a critical factor governing the emergence of property rights is the ability to effectively police use of the claimed object, in terms of being able to enforce duties against nonowners.

17. Nonresident herders destroying trees for fodder might well generate more difficult cases, both in terms of amount of damage and dispute resolution. But again, timely information would not

be the problem it is under the current system and this might reduce unauthorized cutting, particularly since many herders revisit the same villages annually. Controlled barter or sale exchanges of fodder might also develop between herders and tree owners, with trees suffering much less damage as a result. Trees might be pruned, livestock consuming the leaves and villagers stockpiling the stripped branches for fuel wood.

18. See Ingle (1970) for comments on Tanzania's acceding to locally imposed coercion for reasons quite similar to those discussed above.

References

Commons, J. R., 1959. *Legal Foundations of Capitalism*. Madison: University of Wisconsin Press. First published in 1924: New York: Macmillan. Pp. 65-142.

Demsetz, H., 1967. Towards a theory of property rights. *American Economic Review*, 57, pp. 350-53.

Hailey, Lord, 1957. *An African Survey: A Study of Problems Arising in Africa South of the Sahara* (Revised Ed.). Oxford University, London. Pp. 1056-57.

Hardin, G., 1968. The tragedy of the commons. *Science*, 162 (3859): 1243-48.

Higgott, R., and F. Fuglestad, 1975. The 1974 coup d'état in Niger: towards an explanation. *Journal of Modern African Studies*, 13 (3): 383-98.

Ingle, C. R., 1970. Compulsion and rural development in Tanzania. *Canadian Journal of Africa Studies*, 4(1): 77-100.

Le Niger, 1973. Maradi: les eaux et forêts: importante source de développement économique du pays . . . 13(8): 4.

Nicholas, G., 1960. Un village haoussa de la république du Niger: tassad haoussa. *Cahiers D'outre-mer*, 12 (52): 421-50.

Nicholas, G., 1962. Un village Bouzou du Niger: étude d'un terroir. *Cahiers d'outre-mer*, 15 (58): 138-65.

Raeder-Roitzsch, J. E., and F. B. Zenny, 1974. Rapport au Governement de la Haute-Volta sur la Plantification, la Politique et la Legislation Forestières. United Nations, Rome, FAO/PNUD/UPV/71/006: 5, 31, 60.

Raeder-Roitzsch, J. E., 1974. *Institutional Forestry Problems in the Sahelian Region*. United Nations, Secretariat, Special Sahelian Office (ST/SSO/32): 7-9.

Raulin, H., 1960. *Techniques et Bases Socio-Economiques des Sociétés Rurales Nigériennes* ("Etudes Nigériennes"). IFAN, CNRS, n.d., Niamey, Paris, 12: 21-43.

République du Niger, 1971. Projet de développement rural du département de Zinder. Augmentation des resources agricoles des arrondissements de Mirria, Magaria, Matameye. Ministère de L'Economie Rurale Commissariat Général du Développement, Niamey, Niger.

Territoire du Niger, 1957. *Rapport Annuel*, Cercle de Zinder, Subdivision Centrale, Niger, No. 12/C.

Weber, F. R., 1971. *Conservation and Forestry Manual: Niger*. (Prepared for U.S. Peace Corps, no publ.) January 6, pp. 44-45.

5

Rainfall Statistics, Droughts, and Desertification in the Sahel

Richard W. Katz
Michael H. Glantz

The purpose of this paper is to draw attention to the use of the statistics of the climate regime in the West African Sahel and to an underestimation by decision-makers, political and otherwise, of the likelihood of drought in and around arid regions. In these regions where interannual rainfall variability is relatively high, recurrent droughts are in fact a part of climate and not apart from it and are, therefore, not to be viewed as unexpected events (for example, Riehl, 1954, chapter 3; Landsberg, 1975a).

The Sahel has recently been affected by its third major extended drought of this century. Recent reports from different sources have indicated that the drought, which began in various parts of the region in 1968, has ended. Its end has been signaled by what many observers refer to as a return to normal or near normal rainfall. For example, Derek Winstanley (1976), Reginald Newell et al. (1976), and Thomas Dow (1975) each noted that the relatively good rains in 1974 were indicative of a resumption of "normal" rainfall in the Sahel.

Statements such as these have not been uncommon in the growing volume of drought literature. However, they tend to be misleading in that they imply that a return to "normal" rainfall—rainfall seen as an "average"—is associated with a return to acceptable, if not favorable, meteorological, agricultural, and social conditions for the region. Several authors have in fact questioned the validity of such an assumption. Randall Baker (1976, p. 180), writing about the underlying problems in the region, suggested that interest in these problems will dissipate because "when the rains come in the Sahel, and the millet grows again, the 'problem' will be considered over until next time." Michael Mortimore (1973, p. 98), alluding to the problems of development in the Sahel, wrote that "the dangers of using mean rainfall as a basis for development of any kind have been thrown into relief by the present drought."

It becomes apparent that what "normal" rainfall is *perceived* to be can have a major impact, for example, on the sense of urgency associated with the search for solutions to either the persistent underlying regional problems, such as desertification, or to the intermittent problems caused by other natural or human factors, such as those discussed below. Such perceptions also affect the types of development schemes that are planned for the Sahel.

Perceptions of "Normal" Rainfall

One's perception of what constitutes "normal" rainfall is an important, yet often overlooked, aspect of the application of rainfall statistics to the Sahel. Perceived meanings of certain common statistical measures, such as the mean, often differ from their precise mathematical interpretations. These perceptions are extremely important, given the assumption that when one defines a situation as real, that situation will be real in its consequences (Finlay et al., 1967). In other words, people will react to their perceptions whether or not those perceptions are based on reality, and the consequences of those reactions will be real. Climatologist Helmut Landsberg (1975b, p. 1186; 1976a) suggested that the perception of normal has changed since the nineteenth century and that it is misleading to view normal (as it is often viewed) as "a normative quantity like the temperature of the human body, with its connotation that a departure is abnormal and thereby pathological." He also stated: "The notion that a 'normal' has predictive value leads only to confusion. The field of meteorology cannot afford to be thus misunderstood by the public and by other sciences."

The literature concerned with Sahelian droughts is, nevertheless, replete with references to particularly dry years as being abnormal or anomalous. Yet, because perceptions about what constitutes a drought vary, there is no widely accepted definition of drought. The scores of existing definitions, however, could be objectively divided into meteorological and agricultural subgroups (Palmer, 1964; Saarinen, 1966). A meteorological drought could arbitrarily be defined as that time period when the amount of precipitation is less than some designated percentage of the long-term mean. An agricultural drought, on the other hand, could be defined in terms of seasonal vegetation development. If the precipitation is not available to crops such as cereal grains at crucial periods in crop development, the seedlings or, at some later state of crop development, the plant will wither and die.

In reality the subjective distinction between whether a shortfall in precipitation is a dry spell or a drought is a difficult one to make. Commenting on American agriculture, one author noted that most American farmers do not call a dry spell a drought until the moisture shortage has seriously affected the *established* economy of the region (Palmer, 1964).

As an example of the problem associated with making this distinction in the Sahel, one can cite the now famous and increasingly controversial Landsat photographs of the Ekrafane ranch in Niger (figures 5.1–5.3). (The term Landsat [Land Satellite] replaced ERTS [Earth Resources Technology Satellite] in 1975. The change was one of nomenclature only.) To an observer standing in the middle of the ranch, the shortfall of precipitation would have been considered a dry spell. To an observer outside the ranch's perimeter, however, the shortfall would definitely have been considered a drought. One might conclude that with proper range management such as controlled grazing through controlled stock numbers, much of the desert rangelands could have been protected from destruction. In this respect, at least, the human input into the desertification process during extended drought periods may have been considerable.

At a recent workshop on the social implications of a long-range climate forecast, it was suggested that there should be at least two components to a definition of a drought, a physical one and a social one (IFIAS, 1976). A mix of dry spell and meteorological drought occurred at various times and various places in the Sahel between 1968 and 1974, but their impacts were worsened by the pressure of populations, both human and livestock.

An example of the misperception and, therefore, the misapplication of the term drought was referred to by the Paddocks (1967, p. 87), who wrote that

at the slightest absence of rainfall the cry will go forth, "Drought!" "Drought!" "Famine!" "Famine!" The fam-

ine will be genuine but the drought is not a one-time
factor; it is a frequently recurring, normally recurring,
feature of the climate.

Another problem associated with the perception of drought is
the constantly changing nature of those perceptions. It appears
that recent weather occurrences tend to be weighted more heavily
than earlier ones and wet ones more heavily than dry ones. On this,
J. C. Caldwell (1975, p. 17) suggested that

> the [Sahelian] population is likely to compare the dry
> years with the abnormally wet ones. They are likely to
> have adopted practices during the abnormally wet years
> which end in disaster during the dry years.

More specifically related to the most recent Sahelian drought,
Matlock and Cockrum (1976, p. 238) observed that

> by the end of the abnormally high rainfall period, about
> 1965, the expectations of the people had risen to the point

Figure 5.1 (left). Satellite picture showing patch of vegetative debris (lower left) in Sahel.
NASA photo.   Figure 5.2 (center). The patch is the Ekrafane ranch (left) bordered by

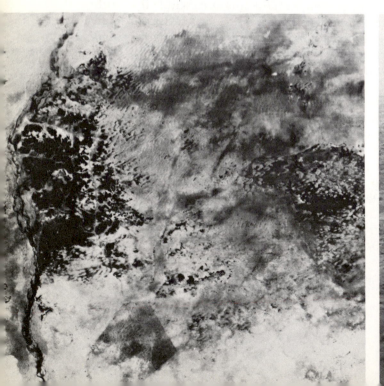

that a return to average conditions was considered to be a drought.

What, then, does a return to average rainfall conditions in arid and semiarid regions mean? Of what importance is it when a return to "normal" or near normal rainfall is perceived to be a return to drought-like conditions? What impact do these perceptions have on development plans for the Sahel? Such questions as these will be discussed in the remaining sections of the chapter. The following section will consider some of the statistical terms which have been used to describe rainfall in the Sahel.

Sahelian Rainfall Statistics

The Sahel has been defined by different authors using variables such as precipitation (e.g., UNESCO, 1975), vegetation (e.g., Stebbing, 1937), and geography (e.g., Tanaka et al., 1975). In this paper it is assumed to be the climatic zone in sub-Saharan Africa which receives 200 to 600 mm of annual mean rainfall. It is

desert (right). Aerial photo by N. H. MacLeod. Figure 5.3 (right). Ground level: the desert is kept out by just a fence and limited grazing inside. Photo by N. H. MacLeod.

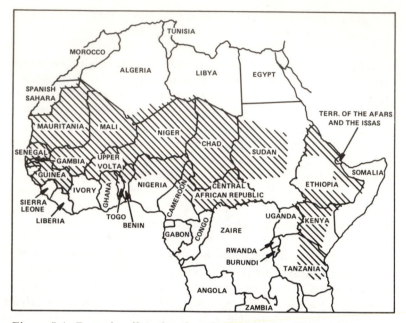

Figure 5.4. Drought affected regions in the Sahel and East Africa, 1970s.

bordered on the north by subdesert (100 to 200 mm) and to the south by the sudan (600 to 900 mm). Rainfall is delivered through convective thunderstorm activities during a four-month period beginning in June (Tanaka et al., 1975). It is this highly variable rainfall on which the region's inhabitants depend for the cultivation of their crops and the maintenance of their herds. The map (figure 5.4) shows those states in the Sahel and in East Africa affected by the recent drought.

The annual and August rainfall records for two cities within the Sahel—Gao, Mali (16.3°N, 0.1°W) and Niamey, Niger (13.5°N, 2.1°E)—will be used in a discussion of rainfall statistics. August rainfall data are significant in that August is nearly always the wettest month and it marks the midpoint of the rainy season. The following graphs (figure 5.6) depict August and annual rainfall observations for both Gao and Niamey, respectively. The observations represent the 1941-75 time period, a period for which adequate rainfall data are available (U.S. Department of Commerce; Bunting et al., 1976).

These graphs illustrate the relatively high degree of variability of Sahelian rainfall. Many analyses, however, do not include a plot

Measures of central tendency

Mean—Arithmetic average; commonly referred to as "average."

Median—Number chosen such that half the observations fall below and half above it.

Mode—Value or range of values which occurs most often.

Running means—Sequence of mean values for each consecutive 5 years (for example) of observations; also called "moving averages."

Measures of variability

Standard deviation—Square root of mean of squared deviations of observations about mean.

Coefficient of variation—Standard deviation divided by mean.

Range—Difference between largest observation (maximum) and smallest observation (minimum).

Measure of shape of distribution

Histogram—Plot of frequency of observations which fall within each of a given set of intervals.

Miscellaneous

Lower quartile—Number chosen such that one quarter of the observations fall below and three quarters above it.

Upper quartile—Number chosen such that three quarters of the observations fall below and one quarter above it.

Quartile range—Difference between upper quartile and lower quartile; also called "interquartile range."

Positively skewed distribution—Distribution for which relatively small observations are more likely than relatively large observations; also called "skewed to the right distribution."

Figure 5.5. Glossary of statistical terms.

of the actual data but have relied on the use of descriptive statistical terms such as the mean or the running mean. Such terms fail to reveal the magnitude of the variability, the awareness of which is a major factor in understanding the nature of rainfall in arid and semiarid lands. Winstanley (1973a), for example, in a detailed analysis of rainfall fluctuations for the Sahel, as well as for other regions, presents only graphs of 5- and 20-year running means. J. Cochemé (1968), too, in a survey of the Sahelian climate, relies on mean monthly and annual rainfall.

Shape of the Distribution

The shape of the distribution of Sahelian rainfall data is important in that interpretations of several statistical terms depend

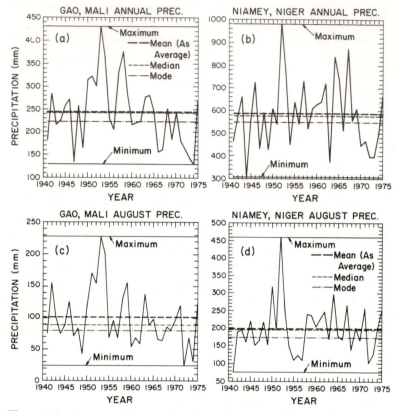

Figure 5.6. August and annual total precipitation with associated statistics for Gao, Mali, and Niamey, Niger, 1941-75.

on it. The distribution of a set of observations is best characterized by a histogram, a chart which portrays the shape of the distribution and which includes information about the central tendency as well as about the variability of the rainfall observations.

The following histograms (figure 5.7), derived from the data for Gao and Niamey, indicate that the frequency distribution of rainfall for each of these stations is asymmetric in that small amounts of rainfall occur more frequently than do relatively large amounts. Such a distribution is said to be "positively skewed."

This type of skewness is characteristic of the distribution of rainfall not only in the Sahel but, more generally, in arid and semiarid regions. In particular, the degree of skewness is greater the drier the climate regime (see values of index of skewness in table

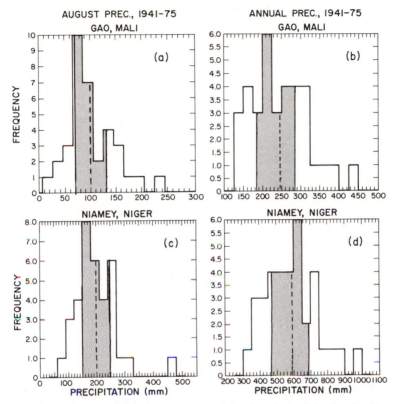

Figure 5.7. Histograms of August and annual total precipitation for Gao, Mali, and Niamey, Niger, 1941-75. Quartile range indicated by shaded area. Left border of shaded area denotes lower quartile; right border denotes upper quartile; dashed line denotes mean.

5.1). As Gao is further north than Niamey, its rainfall distribution is relatively more skewed than is that for Niamey (figure 5.7a as opposed to figure 5.7c; figure 5.7b as opposed to figure 5.7d). In addition, the degree of skewness is greater the shorter the time interval over which the rainfall amounts are totaled (see table 5.1). Thus, August rainfall distributions for both cities are more skewed than are those for annual rainfall (figure 5.7a as opposed to figure 5.7b; figure 5.7c as opposed to figure 5.7d).

It should be added that Landsberg (1975c, p. 196) concluded that the distribution of Sahelian rainfall is actually a mixture of two different rainfall distributions, thus resulting in a bimodal distribution. These two distributions are associated with an

Table 5.1.   Sahelian Precipitation, 1941–75

Statistic	Gao, Mali		Niamey, Niger	
	August	Annual	August	Annual
Sample size	35	35	35	35
Maximum	228.0 mm	431.0 mm	460.0 mm	980.0 mm
Minimum	24.0 mm	129.0 mm	75.3 mm	306.4 mm
Range	204.0 mm	302.0 mm	384.7 mm	673.6 mm
Mean	100.9 mm	243.8 mm	199.7 mm	586.3 mm
Median	89.0 mm	242.0 mm	197.0 mm	577.1 mm
Mode*	80.0 mm	222.8 mm	175.1 mm	548.8 mm
Number in sample less than mean	23	19	19	18
Number in sample less than mode	14	14	15	15
Number in sample within mean $\pm$ 5 mm	4	2	2	0
Number in sample within mean $\pm$ 10 mm	4	3	6	2
Lower quartile	70.0 mm	183.0 mm	151.8 mm	466.0 mm
Upper quartile	132.0 mm	285.0 mm	243.0 mm	690.0 mm
Quartile range	62.0 mm	102.0 mm	91.2 mm	224.0 mm
Index of skewness**	0.456	0.294	0.350	0.253

*Theoretical mode based on fitting gamma distribution to the data.

**Theoretical index of skewness based on fitting gamma distribution to the data.   An index of zero corresponds to no skewness (i.e., symmetry), while the higher the positive value of the index the greater the degree of positive skewness (Kendall and Stuart, 1963, p.85).

alternating circulation pattern: a relatively wet regime exists when the intertropical convergence zone penetrates sufficiently far northward and a relatively dry regime exists otherwise. Landsberg's evidence in support of his conclusion is based on the analysis of rainfall observations for Dakar, Senegal, a coastal West African station for which a relatively long historical record exists. Whether this bimodal distribution also prevails for inland locations within the Sahel has not been established. While there does appear to be a secondary higher mode, indicative of predominance of the intertropical convergence zone, in each of the histograms for Gao and Niamey (figure 5.7), these histograms are not based on enough data to establish whether this secondary maximum is statistically significant.

Central Tendency

Each of the measures of central tendency defined in the glossary (figure 5.5), with the exception of the mode, has been used in one study of the Sahel or another to represent "normal" rainfall. It is, therefore, important to consider how the skewness of the distribution of the rainfall observations affects these measures.

Table 5.1 includes the values of these statistics for the Sahelian rainfall examples, and they are also plotted as part of figure 5.6. In comparing the mean, the median, and the mode, it should be noted that for a positively skewed distribution, the mode is always the smallest of the three and the mean is always the largest. This contrasts with the normal distribution, for which the mean, median, and mode are identical, except for sampling fluctuations. With respect to arid and semiarid zones each of these measures differs in its sensitivity to skewness. Such differences are as follows:

Mean

The mean is extremely sensitive to skewness in that it tends to be inflated by the relatively few cases of large values which cause more than half of the observations to fall below the mean. As an example, 23 of the 35 observations (66%) for Gao August precipitation are below the mean.

In addition, rainfall observations close to the mean value do not occur often in the Sahel. Only 4 of the 35 observations (11%) of Gao August precipitation, for example, fall within 10 mm (10%) of the mean. Hence one might conclude that, at least in some sense, the mean is too large and not at all indicative of how much rainfall commonly occurs in the Sahel, making its interpretation unclear.

Median

The median, as a measure of the midpoint of the distribution of rainfall observations, is affected only by the number of extreme observations and not their values. It is, therefore, less sensitive to skewness than is the mean. In addition, and more important, the presence of skewness does not change its interpretation. The median still marks the amount of precipitation which is exceeded one-half of the time.

Mode

The mode is affected by neither the number nor the value of extreme observations. Thus the mode is not at all sensitive to skewness. It still represents the amount of precipitation which is most likely to occur, and yields the smallest value because there is a high frequency of occurrences of relatively small amounts of precipitation in the Sahel.

The differences among these three measures of central tendency raise the question: which one best represents "normal" with respect to Sahelian rainfall? It can be argued that for the Sahel the mean, in reality, is above "normal," while the median and the mode can still be interpreted, in the sense just discussed, as "normal." Hence it is significant to note that, although most analyses of Sahelian rainfall assume that the mean is an appropriate measure of central tendency, either the median or the mode may, depending on the circumstances, be preferable.

Finally, it becomes apparent that no single number can adequately represent the distribution of Sahelian rainfall, given the large deviations from each of the aforementioned measures of central tendency.

Variability

The standard deviation or coefficient of variation is generally relied upon as the appropriate measure of variability of rainfall in studies of the Sahel.

Standard Deviation

The standard deviation is a statistic which only measures the variability well when the observations are normally distributed or at least symmetrically distributed with few extreme values. Because it is a measure of the degree to which each observation deviates from the mean, the standard deviation is based on a measure of central tendency which, as has been noted for Sahelian rainfall data, is too large.

The standard deviation is overly sensitive to extreme values because they result in large, unrepresentative squared deviations from the mean. Consequently, a few relatively large amounts of precipitation may make an inordinate contribution to the standard deviation.

Because the coefficient of variation differs from the standard deviation only in division by the mean, it has the same deficiencies. These deficiencies obscure the interpretation of the standard deviation and coefficient of variation, making them inappropriate for characterizing the variability of Sahelian rainfall.

Range

The range is a simple measure of variability. It is not particularly sensitive to skewness because it does not depend on the shape of the distribution curve. Its interpretation is, therefore, always clear; it is the difference between the greatest and the smallest amounts of rainfall ever observed. The values of the range for the Sahelian examples (see table 5.1), when compared to any of the measures of central tendency, show the relatively wide variability of precipitation in this region, for example a range of 204.0 mm (24.0 mm to 228.0 mm) for Gao August rainfall.

Quartile Range

The quartile range is a variant of the range which provides information about both the central tendency and the variability of the data. Because it gives the range of the middle 50% of the observations, it can characterize the central tendency more adequately than can any single number. In addition, the quartile range conveys more information about the central tendency than the ordinary range since it excludes the extreme observations. Thus one might argue that the middle 50% of the observations, for example 70.0 mm to 132.0 mm for Gao August rainfall, gives a better idea of the notion of "normal" rainfall than any of the standard measures of central tendency or variability (see table 5.1 and figure 5.7).

This suggests that there may be simple yet relatively more meaningful alternatives to the standard statistics used to describe Sahelian rainfall.

Trends

Measuring the variability of Sahelian rainfall is further complicated by the question of whether or not there are trends or cycles or other evidence of climatic change in the data. Discussion of rainfall trends in the Sahel has generated much controversy, with a substantial amount of the scientific literature on the Sahel being concerned at least tangentially with this issue.

Winstanley (1973b, p. 465), in analyzing the patterns of running means for Sahelian precipitation, on the one hand, has suggested that there has been a decline in rainfall since the 1930s, "a trend which was first evident in the 1930s, but which became even more marked in the 1960s." Helmut Landsberg (1975c), too, on the basis of a spectral analysis of Dakar, Senegal, rainfall data, has observed cycles of 2- to 3-year and 11-year periodicity.

On the other hand, A. H. Bunting and his colleagues (1976, p. 59) have claimed that "no established trends or periodicities can be detected, and the recent succession of drought years in the Sahel falls within statistical expectation." In a more general context not specifically related to the Sahel, V. Yevjevich (1974, p. 228) has lamented that "the 60 to 70 years of hidden periodicity research has produced several thousand references, but no clear periodicity beyond a year could be definitely confirmed for hydrologic processes."

This does not mean that such periods might not exist, but it does suggest that there may have been an inordinate degree of effort devoted to "hidden periodicity" research. The debate about the existence of rainfall trends and cycles has tended to obscure the most important aspect of Sahelian rainfall, its extreme variability. In fact, even if trends or cycles do exist, it is almost universally agreed that they would be of limited predictive value. In questioning the usefulness of the search for trends, Landsberg (1975b, p. 1186) has stated:

> Physical models are not yet discovered for many of the fluctuations and iterations of climate, but probably do exist. A "trend," on the other hand, is a statistical fiction without predictive validity; it has no predictive value whatever, and we have no idea when it will change. This word seems to have induced as much recent confusion as "normal" did.

Other Factors Affecting the Social Value of Rainfall

Aside from the absolute amounts of rainfall, as measured over varying lengths of time, other factors such as spatial distribution, timing, wind velocity, and evapotranspiration could have an important impact on the actual social value to the inhabitants of the rain that does fall. For example, the spatial distribution in any given season is extremely variable. On this J. C. Caldwell (1975, p. 19) has written:

the records of the main meteorological stations in Niger show that in the drought year of 1968 Agadez and Tahoua received above average [mean] rainfall; in 1969 Iferouane recorded one and a half times its average [mean] but Agadez only half its usual amount; in 1970 Tahoua was once again above average [mean].

Michael Horowitz (1972, p. 106), referring to some problems of rainfall timing, noted:

The [rain]fall may be evenly distributed throughout the four months of the rainy season or may be concentrated in a few intense periods. Contiguous areas often experience much difference. . . .

In 1968 and 1969 [for example] early rains led the grass to germinate, and the immature seedlings were subsequently burned by the sun.

As a final example of the importance of other rainfall-related forecasts, Caldwell (1975, p. 3) noted that " 'rainfall' figures are, by temperature standards, misleading, as they are low compared with the levels of evapotranspiration."

In summary, one might conclude that if rainfall information, which represents only one climatic variable, is not viewed within its total environmental context, it can be extremely misleading. Where long-range development planning is concerned, such an oversight can be disastrous. Such has often been the case in fragile ecosystems like those in arid and semiarid regions. In fact, improper use of rainfall statistics with respect to development planning efforts can exacerbate the desertification process, defined in this paper as the creation of desert-like conditions where none had existed before.

Desertification

Desertification is the result of either the vagaries of weather and climate or the mismanagement of the land or, as in most cases, some combination of both.

Ecological deterioration in the Sahel has been linked in several ways to livestock herds. The introduction of modern medicines and medical services into the pastoral system, however intermittent, has served to keep larger numbers of livestock and humans alive (Lambrecht, 1972). In times perceived by government officials and other observers to be normal, that is, during periods of above-average rainfall, herders and cultivators had a symbiotic relationship. The pastoralists, who would sometimes herd the cattle of the cultivators along with their own, would be allowed to graze their cattle in the fields following the harvest and, in return, cattle manure would be used to fertilize the fields for the next crop year. In times of favorable rainfall the region in which the pastoralists lived was being taken from them as the cultivators, supported by government policies, pushed farther north into more marginal semiarid and arid areas in order to put more land into the cultivation of cash crops. According to a recent U.N. FAO report (1974, p. 3), the present situation in the Sahel has been

> due to the buildup over the years before the drought, of high human (and animal) populations relative to the carrying capacity of the land. This trend has been magnified by even greater population increases outside the range area, leading to an expansion of cultivation and hence, a reduction of available grazing area.

Some of the marginal land that had been cleared for cultivation was later abandoned because of low crop yields, and was thereby left open to wind erosion and desiccation. Pastoral populations became sandwiched on increasingly marginal land on the southern edge of the Sahara, between the desert to the north and the cultivators (as well as the tsetse fly) to the south. During extremely favorable periods of rainfall, the pastoralists inhabited this marginal region with relatively large herds. However, with the onset of a series of dry years (drought) beginning at the end of the rainy season in 1967, the pastoral populations found themselves occupying very marginal rangelands whose carrying capacity was extremely overtaxed by the large number of livestock. The end result was that the nomads who had been pushed into these marginal areas along with the sedentary farmers who sought to open up new, yet marginal, lands to cultivation viewed themselves as victims of a natural disaster when the unusually high rainfalls ended. The recent reduction in available grazing area due to drought coincided with the time when expanded herds were in need of greater availability of forage and water. Other reports similar to the one issued by the FAO have affirmed the view that the periodic droughts which have plagued the region have only tended to exacerbate a situation in which ecological deterioration was already well in progress. As an example, it has been noted that existing data on the Malian herds suggested that the stocking rate in relation to the carrying capacity had been exceeded before the onset of the shortfall in precipitation in 1968 (Clyburn, 1974).

An understanding and use of statistical measures by government officials, for example, could have maintained an awareness of expected rainfall variations. Summarizing this situation, one author (Ware, 1975, p. 35) noted:

> The result is that after a series of good years the distribution of human populations and the cattle numbers are not such as can be maintained in less favorable years.

Playing the Odds

A major step toward combatting deterioration, and ultimately desertification, in the Sahel and in similar ecosystems would be the determination of an ecologically realistic carrying capacity for the rangelands. Although there appears to be widespread acceptance that such a determination would be a necessary, but not in itself sufficient, step toward the conservation of the rangelands, there has been little agreement to date on how to make operational the notion of carrying capacity, which is defined as "the amount of grazing stock that the pasture can support without deterioration of either the pasture or the stock" (Boudet, 1975).

Should the carrying capacity be geared to the best, the average, or the poorest years? Which combination of statistical measures would be most meaningful for the planning of long-term development of the rangelands? On which variables should such an assessment be based—vegetation, rainfall, soil, ground and surface water, or managerial capabilities? Such inconclusiveness within the scientific community, while understandable, tends to create confusion for the decision-makers, political and otherwise, in the Sahelian and in the donor states. Often, they decide either to take no action at all or they decide that all scientific suggestions are of equal weight and, therefore, indiscriminately choose any one of those suggested. Given the downward spiral of land deterioration,

drought → less than perfect recovery → drought → and so on,

it becomes essential that an ecologically acceptable carrying capacity be established and enforced. Such planning can be achieved only with an understanding of the statistical properties of Sahelian rainfall.

It would be neither difficult nor misleading to argue that a lack of understanding of these statistical properties has in part contributed to desertification in general and more specifically to the deterioration of the rangelands. The use of terms such as *normal* and *average* has been extremely misleading and must be qualified when used. Other relatively more informative and therefore more meaningful terms such as *mode, quartile,* or *range* would more adequtely represent rainfall patterns of arid and semiarid regions. In addition, the use of such terms would tend to develop a higher level of awareness of actual Sahelian rainfall conditions rather than those conditions which observers believe, or would like to see, prevail, such as mean rainfall.

A distinction between the real nature of climate and how it is often perceived is an important one to make. The mistaken idea, for example, that drought in the Sahel is an unexpected event has often been used to excuse the fact that long-range planning has failed to take rainfall variability into account. Focusing attention on this problem, Helmut Landsberg (1976b) has noted:

> I wish people would quit blaming climate or climate change for failure of agriculture in marginal lands. It's a scapegoat for faulty population and agricultural policies.

Jean Cochemé (1968, p. 236) touched on the importance to development planning of being familiar with the climatic conditions in the Sahel, when he wrote that

> local traditional agriculture may have learnt by dire experience to cope approximately with the climate of the area, but a sound knowledge of climatic resources will save time and money not to mention the ecosystem in the choice and implementation of coordinated changes.

As applied to social conditions, one might ask, "What is normal in the Sahel?" On this point one journalist (Kamm, 1974) recently observed:

> For the villagers the end of drought, if the rains of this autumn [1974] return next year and the year after, means a return to "normal" life—on the edge of hunger, consistently malnourished, beset by malaria and hookworm and other parasites, having average life spans that fall short of forty years.

Recall that four of the six Sahelian states are literally landlocked while a fifth, Mauritania, is functionally so. They are listed among the poorest in the world with per capita incomes less than $100 per year and with food production falling behind population growth. These states have weak infrastructures and, as an example, Niger and Chad have no national rail system. Malnutrition is widespread in the Sahel as well as in other parts of Africa, although the levels vary between countries, between regions, and between ethnic groups. In this respect, therefore, a return to better rainfall means a return to a lesser level of malnutrition; that is, a continuation of malnutrition but at a more acceptable level. During those rainfall periods which most frequently occur—periods of relatively low rainfall—malnutrition problems faced by the aged and the young are compounded (Kloth, 1974, p. 1):

> During times of food scarcity, certain customs of West Africa culture may dictate that children receive less than their proportionate share of food.

Therefore, the effects of nutritional deprivation are a greater risk to the children than to the adults. Normal, when applied to social conditions, also means a return to precrisis political and social cleavages within the Sahel as well as within the Sahelian states: government versus nomadic populations, pastoralists versus sedentary farmers, recipients versus donors, recipients versus recipients.

Notions such as normal, if used, must be qualified in such a way that the reader will be better informed about the total geographical conditions of the region to which the notion has been applied. Landsberg (1976c, p. 213), taking a more radical approach, has even suggested that the term normal "should be expurgated from the vocabulary when talking about climate." The use of normal, however, will continue. Therefore, it will be crucial that decision-makers, political, economic, and otherwise, especially at the national and international levels, know what these statistical and quasi-statistical measures actually mean. As emphasized earlier, no single number can adequately describe the climate regime of an arid or semiarid region. Decision-makers must supplement such terms as the *mean* with more informative statistical measures in order to adequately characterize the variability of the Sahelian climate. An awareness of this high degree of variability will serve to remove one of the major obstacles to resolving the perennial problems of the Sahel.

References

Baker, R., 1976. Innovation, technology transfer and nomadic pastoral societies. In M. H. Glantz, ed., *The Politics of a Natural Disaster: The Case of the Sahel Drought*. New York: Praeger Publishers.

Boudet, G., 1975. Problems posed in the estimation of the rate of stocking of natural pastureland. Presented at International Livestock Center for Africa Colloquium on the Mapping and Evaluation of Rangelands in Tropical Africa (March 3-8), Bamako, Mali.

Bunting, A. H., M. D. Dennett, J. Elston, and J. R. Milford, 1976. Rainfall trends in the West African Sahel. *Quarterly Journal of the Royal Meteorological Society, 102*: 59-64.

Caldwell, J. C., 1975. *The Sahelian Drought and Its Demographic Implications*. American Council on Education, Overseas Liaison Committee Paper No. 8, Washington, D.C.

Clyburn, L., 1974. A commentary on the FAC outline of the livestock development strategy for the Sahel States. Mimeo draft, USAID/CWR, Washington, D.C.

Cochemé, J., 1968. Agroclimatology survey of a semiarid area in West Africa south of the Sahara. In *Agroclimatological Methods: Proceedings of the Reading Symposium*. UNESCO, Paris.

Dow, T. E., 1975. Famine in the Sahel: a dilemma for United States aid. *Current History, 68*: 197-201.

Finlay, D. J., D. R. Holsti, and R. R. Fagen, 1967. *Enemies in Politics*. Skokie, Ill.: Rand McNally & Co.

Horowitz, M., 1972. Ethnic boundary maintenance among pastoralists and farmers in the Western Sudan (Niger). *Journal of Asian and African Studies, 7*: 105-114.

IFIAS (International Federation of Institutes for Advanced Study), 1976. Workshop on the Social Implications of a Credible and Reliable Climate Forecast (Jan. 26-29), Aspen Institute, Berlin.

Kamm, H., 1974. Sub-Saharan lands are hopeful of a reasonable harvest soon. *New York Times*, 10 November.

Kendall, M. G., and A. Stuart, 1963. *The Advanced Theory of Statistics V. 1* (second edition). New York: Hafner.

Kloth, T. I., 1974. *Sahel Nutrition Survey, 1974*. U.S. Public Health Service, Center for Disease Control, Atlanta.

Lambrecht, F. L., 1972. The tsetse fly: a blessing or a curse? In M. T. Farvar and J. P. Milton, eds., *The Careless Technology*. New York: Doubleday & Co. Pp. 726-741.

Landsberg, H. E., 1975a. Drought, a recurrent element of climate. In *Drought*. WMO Special Environment Report No. 5, Geneva, pp. 45-90.

Landsberg, H. E., 1975b. Quoted in D. H. Miller, Symposium summary—a return to skepticism. *Bulletin of the American Meteorological Society, 56*: 1185-1186.

Landsberg, H. E., 1975c. Sahel drought: change of climate or part of climate? *Arch. Met. Geoph. Biokl., Ser. B, 23*: 193-200.

Landsberg, H. E., 1976a. The definition and determination of climatic change, fluctuations and outlooks. In R. J. Kopec, ed., *Atmospheric Quality and Climatic Change*. University of North Carolina, Department of Geography, Studies in Geography No. 9.

Landsberg, H. E., 1976b. Personal communication to M. H. Glantz.

Landsberg, H. E., 1976c. Concerning possible effects of air pollution on climate. *Bulletin of the American Meteorological Society, 57*: 213-215.

Matlock, W. G., and E. L. Cockrum, 1976. Agricultural production systems in the Sahel. In M. H. Glantz, ed., *The Politics of a Natural Disaster: The Case of the Sahel Drought*. New York: Praeger Publishers.

Mortimore, M., 1973. Drought in Africa: introduction. *Savanna, 2*: 97-101.

Newell, R. E., M. Tanaka, and B. Misra, 1976. Climate and food workshop: a report. *Bulletin of the American Meteorological Society, 57*: 192-198.

Paddock, W., and P. Paddock, 1967. *Famine 1975!* Boston: Little, Brown and Co.

Palmer, W. C., 1964. *Meteorological Drought.* U.S. Weather Bureau, Department of Commerce Research Paper No. 45.

Riehl, H., 1954. *Tropical Meteorology.* New York: McGraw-Hill Book Co.

Saarinen, T. F., 1966. *Perception of Drought Hazard in the Great Plains.* University of Chicago, Department of Geography Research Paper No. 106.

Stebbing, E. P., 1937. The threat of the Sahara. *Journal of the Royal African Society, 36* (Supplement): 3-35.

Tanaka, M., B. C. Weare, A. R. Navato, and R. E. Newell, 1975. Recent African rainfall patterns. *Nature, 255*: 201-203.

UNESCO, 1975. Regional meeting on integrated ecological research and training needs in the Sahelian region. Man and the Biosphere Program, Paris.

UN/FAO, 1974. The ecological management of arid and semiarid rangelands in Africa and the Near East: an international programme. FAO Publication AGPC: MISC/26, Rome.

U.S. Department of Commerce,. *Climatic Data for the World,* Washington, D.C. (published monthly).

Ware, H., 1975. The Sahelian drought: some thoughts of the future. U.N. Special Sahelian Office, New York.

Winstanley, D., 1973a. Rainfall patterns and general atmospheric circulation. *Nature, 245*: 190-194.

Winstanley, D., 1973b. Recent rainfall trends in Africa, the Middle East and India. *Nature, 243*: 464-465.

Winstanley, D., 1976. Climate changes and the future of the Sahel. In M. H. Glantz, ed., *The Politics of a Natural Disaster: The Case of the Sahel Drought.* New York: Praeger Publishers. Pp. 189-213.

Yevjevich, V., 1974. Determinism and stochasticity in hydrology. *Journal of Hydrology, 22*: 225-238.

6

Ocean Deserts and Ocean Oases

J. Dana Thompson

A book on deserts seems an unlikely place to find a discussion of the world's oceans. Yet most of the oceans are considered to be deserts in a biological sense. Approximately 90% of the world's ocean areas, nearly three-fourths of the earth's surface, produce a negligible fraction of the world's fish catch at present, and have little natural potential for yielding more in the future (Ryther, 1969). Furthermore, biologically productive ocean "oases" often border terrestrial deserts, since the climatic conditions favorable for above-average ocean productivity are often important factors in the creation of coastal deserts. The most productive fishery of the world's oceans, situated within 100 km of the Peruvian and Chilean coasts, is adjacent to one of the most arid deserts on earth (Trewartha, 1961).

In this chapter we will examine the notion of biological deserts in the ocean and assess the limitations of the sea as a biological resource. We will outline the evidence for ocean deserts and ocean oases and discuss the physical mechanisms responsible for their existence. The connection between coastal deserts and certain productive coastal fisheries will be explored. Finally, we will investigate the importance of natural climatic and biological variability, marine pollution, and overfishing on the process of biological "desertification" in the ocean.

Ocean Deserts and Ocean Oases: The Biological Evidence

The potential food value of the sea has become a topic of increasing interest in recent years (National Academy of Sciences, 1975). A growing body of evidence suggests that the oceans are not the unlimited sources of food as once supposed. While we still know very little about marine productivity (Riley, 1972), recent estimates of maximum sustainable fish yields in the ocean are only two or three times greater than present fish harvests. Further, some historically productive fisheries have declined in recent years and others are experiencing severe overfishing. The problem of ocean "desertification," here defined as significant reduction of marine biological productivity useful to man, is both serious and in need of study. To provide some perspective to the problem, it is useful to examine first the theoretical and practical limits to marine productivity and to differentiate between biologically productive and unproductive areas in the world's oceans.

Most estimates of the oceans' potential food value to man are based on photosynthetic (primary) organic production rates. They assume a certain number of "links" or trophic levels in the food chain between primary producers (phytoplankton) and man, and certain efficiencies of organic conversion from level to level. These assumptions lead to varying estimates for ocean food potential. While such estimates have large uncertainties and are rather controversial (Paulik, 1971), they do provide a necessary background for our discussions.

About 0.2% of the solar energy incident on the earth's surface each year is utilized for photosynthetic processes, which produce a total each year of about 6×10^{10} dry-weight metric tn of organic matter (Hela, 1971). Roughly one-third is produced in the oceans, two-thirds on land (UNESCO, 1970). Per unit area the land is, on the average, over four times more productive than the oceans. If we assume a three-trophic-level system (phytoplankton feed zooplankton feed fish), a 90% organic conversion loss from level to level, and full utilization on each trophic level (for example, all zooplankton are eaten by fish), then the oceans could theoretically produce nearly 2 wet-weight kg of fish per person in the world per day (Hela, 1971). By contrast, in 1974 the world's commercial catch of fish, crustaceans, mollusks, and all other aquatic plants and animals (except whales and seals) was about 69.8 million metric tons (mmt) (FAO, 1974), or only 3.5% of the theoretical maximum for fish alone. Clearly, theoretical fish production is not the potential, much less the actual, fish harvest. Many factors have been omitted in our estimate of available fish. Man must compete with other carnivores—birds, sea lions, etc.— for the same fish. He must leave a substantial fraction of the fish to provide future sustainable yields. Many fish species do not congregate in economically "catchable" schools. The net result is that maximum sustainable yield estimates are drastically lower than the above calculations would indicate. Recent estimates (Hood and McRoy, 1971; Ryther, 1969; Cushing, 1969) of maximum sustainable yields from hunted fish (contrasted to "farmed" fish in aquaculture) range from 100 to 200 mmt per year, or little more than twice present yields.

Although one might still be optimistic that a doubling of fish yields could be realized by expanded fisheries, the facts are not so encouraging. One reason is that oceanic productivity is highly concentrated in certain areas which are already near, at, and in some cases, over, maximum sustainable fish yields. This heterogeneous distribution of ocean productivity attests to the existence of ocean deserts and ocean oases.

Plants in the sea, as on land, require light, nutrients, trace metals, and a suitable medium in which to grow. While sea plants are subjected to less extreme temperature fluctuations than land plants, they are more susceptible to fluctuations of nutrients. Given sufficient light, the availability of nutrients is generally the most important factor in phytoplankton growth, and they are said to be "nutrient-limited." In the upper 100 m or so of the ocean, in the sunlit or euphotic zone, nutrients are assimilated by the phytoplankton, which grow, die, and sink out of the zone, carrying their nutrients with them. If those nutrients are not replaced, large phytoplankton populations cannot be maintained. While land runoff may provide some nutrients for phytoplankton growth near coastal margins, it is largely upward vertical currents (upwelling) and turbulent mixing which return decomposed organic material and nutrients into the euphotic zone.

A map showing the areas of strong upwelling in the world's oceans also adequately serves as a map of the areas of high organic productivity (Smith, 1968). In most of the open seas of the world, away from continental margins, vertical currents are relatively weak; consequently annual primary production is low. Ryther (1969) has estimated that in 90% of the oceans the average primary production, quantitatively measured in terms of the amount of inorganic carbon converted into organic carbon by photosynthesis ("fixed") per unit of ocean area per unit of time, is only about 50 g per m² per year, with a range of 25 to 75 g per m² per year.

Roughly 10% of the world's ocean areas have a moderate rate of primary production: 100 g of carbon fixed per m² per year. They include those shallow coastal zones (waters less than 180 m deep) where strong upwelling currents are absent but where nutrients are provided by land runoff and by vertical mixing down to the shallow ocean bottom. Also included are open ocean areas where dynamical conditions produce vigorous upwelling, as in oceanic fronts and in equatorial divergences.

Finally, there are the coastal upwelling areas, which comprise about 0.1% of the total ocean area (figure 6.1). According to Ryther (1969):

> They exist off Peru, California, northwest and southwest Africa, Somalia, and the Arabian coast, and in other more localized situations. Extensive coastal upwelling also is known to occur in various places around the continent of Antarctica, although its exact location and extent have not been well documented. During periods of active upwelling, primary production normally exceeds 1.0 and may exceed 10.0 grams of carbon per square meter per day. Some of the high values which have been reported from these locations are 3.9 grams for the southwest coast of Africa (Nielsen and Jensen, 1957), 6.4 for the Arabian Sea (Ryther and Menzel, 1965) and 11.2 off Peru (Ryther, et al., 1969). However, the upwelling of subsurface water does not persist throughout the year in many of these places—for example, in the Arabian Sea, where the process is seasonal and related to the monsoon winds. In the Antarctic, high production is limited by solar radiation during half the year. For all these areas of coastal upwelling throughout the year, it is probably safe, if somewhat conservative, to assign an annual value of 300 grams of carbon per square meter.

Despite this high primary production rate, quantitative correlations between upwelling and fish production are not easily demonstrated. The cold upwelling water may temporarily inhibit phytoplankton growth. Time elapses between phytoplankton blooms and zooplankton growth and between zooplankton growth and fish production. Currents may carry the constituents of each

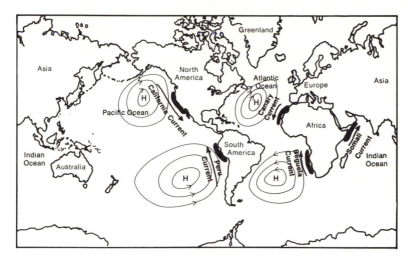

Figure 6.1. Major coastal upwelling regions of the world and the sea-level atmospheric pressure systems (anticyclones) that influence them.

trophic level out of the region of intense upwelling. Fish may move out on their own.

In the three ocean regimes, conversion efficiencies and the number of trophic levels between phytoplankton production and fish production vary. Ryther (1969), Cushing (1969), Gulland (1968), and others have attempted to relate potential fish production to primary production using educated guesses as to the trophic levels and conversion efficiencies involved. The following table (6.1) is based on Ryther's 1969 estimates. The table underscores two important characteristics of coastal upwelling ecosystems. First, fewer trophic levels are involved than in open ocean or coastal regimes. Man can harvest animals from trophic levels very near the primary producers. Second, the conversion efficiencies are also higher than the other two regimes, although these estimates have considerable uncertainty. Based on these estimates, we calculate that coastal upwelling zones are about 66,000 times more productive per unit area than the open ocean. Is it so surprising then that less than 0.1% of the world's ocean areas yield half the world's food catch? Such a distribution of fish production in the sea is dramatic biological evidence for the existence of ocean deserts and ocean oases.

Table 6.1. Estimated Fish Production in the World's Oceans
            (Based on estimates from Ryther, 1969)

Regime (% Ocean Area)	Trophic Levels	Efficiency %	Fish Production (Fresh Wt. Tons Per Year)	Fish Production Per Unit Area Per Year (Relative to the Open Ocean)
Oceanic (90)	5	10	16×10^5	1
Coastal (9.9)	3	15	12×10^7	660
Upwelling (.1)	1.5	20	12×10^7	66,000

Despite the uncertainty in our estimates, we are forced to conclude that the open oceans are indeed biological deserts when compared to the narrow, productive coastal upwelling regions. Unfortunately, these zones of highest ocean productivity are closest to man's sources of pollution and his technological tools for species decimation.

Upwelling, Climate, and Coastal Deserts

Some evidence for biological deserts and oases in the oceans has been presented. It has been suggested that ocean productivity is closely linked to vertical currents and mixing of the upper ocean. As yet, the physical mechanisms responsible for oceanic upwelling have not been identified.

Open Ocean Upwelling and Downwelling

The surface wind systems of the world are largely responsible for driving the circulation of the upper ocean. (For our purposes, "upper ocean" will refer to the upper 500 m of the water column.) Upwelling and marine productivity are significantly influenced by these wind patterns. In this brief chapter we cannot hope to offer a thorough review of the wind-driven ocean circulation or its influence on marine productivity. (For such a review, see Defant, 1961.) Fundamental physical mechanisms important to our discussion can be briefly outlined using the following highly idealized physical model.

Centers of high atmospheric pressure at sea level are generally situated at subtropical latitudes over the world's oceans. In the absence of the earth's rotation, we might expect surface winds near these centers to blow outward from high to low pressure, perpendicular to lines of constant pressure (or isobars). Due to the earth's rotation, these winds tend to turn to the right in the Northern Hemisphere and to the left in the Southern Hemisphere, aligning themselves roughly parallel to the isobars. Observe, for example, that the large North Pacific high, situated off the west coast of North America near 30°N and 150°W in summer, is characterized by surface winds extending several thousand km from the high center, blowing clockwise around it. Due to frictional drag at the sea surface, the winds are directed at an angle slightly outward from the isobars (see figure 6.1).

Surface currents in the ocean are blown along by these winds, and, in the absence of the earth's rotation, would move in the same direction as the winds. This rotation is of considerable influence, even to ocean currents, and the surface waters tend to be driven to the right of the wind direction. In fact, a careful mathematical analysis by V. Walfrid Ekman (1905) showed that the net effect of winds driving ocean surface waters on a rotating earth was to produce a net transport of water (later termed "Ekman Drift") directed 90° to the right (left) of the wind in the Northern (Southern) Hemisphere. Under a surface high, the current has a component directed inward toward the center and the waters begin to pile up, tending to raise the sea level under the high.

Clearly, surface water cannot converge indefinitely under the high's center. Instead, it must move downward, then outward at some lower level, then upward again far from the high center to provide mass balance. Downwelling is thus established in the upper ocean under a surface high. Generally this downwelling is persistent, encompasses thousands of square kilometers of ocean area, and is quite weak, only a meter or so in a month. Nevertheless, it inhibits the vertical transport of nutrients into the upper ocean and helps explain the desert-like primary productivity of the open ocean. Some vertical exchanges of water and nutrients still occur, particularly during the fall season when the surface waters become cooler than subsurface waters and grow convectively unstable, producing turbulent vertical exchange in the upper ocean. Consequently, these open ocean regions are not totally devoid of plant life.

Occasionally a cyclonic low pressure system such as a hurricane will produce a strong diverging surface current over the open ocean. Surface water will spiral outward from the low center and a vigorous upwelling of several meters per day will result. Since these systems usually move rapidly and are short-lived, they do not substantially increase the nutrient content of the upper ocean.

One region of the open ocean where persistent, vigorous upwellings are present is within a few degrees of the equator. The equatorial perimeter of the circulation around the oceanic highs marks a zone of fairly steady easterly trade winds. Precisely on the equator, the direct influence of the earth's rotation is absent and winds blowing westward tend to drive westward surface currents. Just to the north of the equator the earth's rotation is felt by the currents and they tend to turn to the right of the wind. Just to the south, they turn to the left. The net result is a divergence of surface water away from the equator, with replacement water coming from a few hundred meters below the sea surface. Nutrients and cooler subsurface waters are lifted up into the euphotic (sunlit) zone and lead to increased marine productivity.[1] As noted earlier, equatorial divergence zones may have primary production rates as high as some coastal areas. These areas have received increased study, both observationally and mathematically, in recent years (Wyrtki, 1966; Moore and Philander, 1976).

Coastal Upwelling

The eastern flanks of the subtropical oceanic highs are generally situated near the west coasts of continents, as shown in figure 6.1. This is not simply coincidental, since land-sea contrasts of temperature and surface topography influence the wind systems. An extreme example is the low-level atmospheric winds off western Peru, which tend to be channeled along the coast by the high coastal range of the Andes. Even in the absence of terrain features, there is generally some time during the year at mid-latitudes when the predominant surface winds blow equatorward along the western coastline of continents. The Ekman theory, as extended by Thorade (1909) and Sverdrup (1938), predicts an offshore turning of the coastal surface currents to the right (left) of the wind in the Northern (Southern) Hemisphere, and a net offshore transport of surface water. To replace the diverging surface water, there must be compensating subsurface currents and upward vertical motion, or coastal upwelling. Formally, coastal upwelling may be defined as an ascending motion of some minimum duration and extent by which water from subsurface layers is brought into the surface near the coast and removed by horizontal flow (Smith, 1968). Recent theoretical and observational studies suggest that not all coastal upwelling is locally wind-driven. Continental shelf waves and internal Kelvin waves have been shown to influence coastal upwelling circulations and are presently an important avenue of research in physical oceanography (Allen, 1976; Hurlburt and Thompson, 1976; Houghton and Beer, 1976).

Despite evidence that some coastal upwelling is not driven locally by the wind, the Ekman-Sverdrup theory has led to a useful index for coastal upwelling—the Ekman transport away from the coastal boundary. Its magnitude is proportional to the surface wind stress divided by the sine of the latitude. This index has been used by Wooster and Reid (1963), Bakun (1973), and others to identify regions of potentially strong coastal upwelling. In the Wooster and Reid study, seasonal estimates of Ekman transport of 5° intervals of latitude along west coasts of continents were calculated. Smith (1968) has summarized their findings below (maximum values of the index suggest maximum coastal upwelling rates):

(1) The maximum values are usually observed in the spring or summer, with the exception of the distinct winter maximum for the Peru Current region north of 30°S.

(2) The maximum values of the index migrate from south to north, from spring to summer, in the California, Benguela, and Canary Current regions.

(3) The maximum values for the index are in the Southern Hemisphere. The smallest values are off the west coast of North America.

(4) Negative values are observed at high latitudes (poleward of 50°).

One region not examined in their study was the western Indian Ocean, along the Arabian and Somali coasts. During the Southwest Monsoon, winds blowing parallel to the coast produce extremely vigorous upwelling. Also, though their index predicts upwelling off the west coast of Australia, observations do not show it to be significant.

Recently the physical oceanography and biology of coastal upwelling areas have received widespread attention. Elaborate field studies off the coasts of the United States (CUE-I, II, 1972 and 1973), northwest Africa (JOINT I, 1974), and Peru (JOINT II, 1975-1977) have been mounted to study coastal upwelling circulation dynamics, biological productivity, and marine meteorology. The United States' effort has largely been coordinated by the Coastal Upwelling Ecosystem Analysis (CUEA) Program, funded by the National Science Foundation within the International Decade for Ocean Exploration. One result of these studies is that our understanding of coastal upwelling, its length and time scales and the complex processes within the food chain, has been greatly increased during the last five years. As a prelude to the remainder of this chapter, it is worthwhile to review a few basic facts concerning the primary coastal upwelling areas of the world.

As previously discussed and as shown in figure 6.1, there are five primary coastal upwelling regions: off the west coast of South America (primarily Peru), northwest Africa, southwest Africa, the east coast of Africa from Somalia northward into the Arabian Sea, and the western coast of North America from Baja to the Canadian border. They all have identifiable coastal current systems within several hundred kilometers of their coastlines. There is usually one season in which coastal upwelling predominates. During that season, sea surface temperatures within 100 km of the coast may be several degrees centigrade colder than offshore waters. This tends to suppress atmospheric convection and helps stabilize the marine boundary layer. As a result, the coastal climates along upwelling zones tend to be cooler, more arid and, somewhat paradoxically, more humid than inland areas at the same latitude. Each area is notable for its fishery, Peru being the most productive. Despite the many general similarities, each area is unique in the details and requires closer examination.

Peru. The most famous coastal upwelling region in the world lies in an area roughly 800 x 30 mi (Wooster and Reid, 1963) off the Peruvian coast. During the seven-month fishing season of 1969-1970, 11 mmt of a single fish species, *Engraulis ringens* (the Peruvian anchoveta), were harvested in that region (Paulik, 1971). The total U.S. catch for all species of fish and shellfish in 1969 was about 2.5 mmt. Although Peru's catch that year could have provided about 1 tn of high quality protein to each of its citizens, much of it was actually exported in the form of fish meal to the industrialized nations of Western Europe and North America to be used as poultry, swine, and cattle feed. The story behind Peru's fishing boom (and recent apparent decline) is worthy of a Steinbeck novel. However, our focus here is on the physical environment.

The dominant feature of the physical oceanography along the 1,475-mi Peruvian coastline is the Peru Current. Since the first major oceanographic survey of the region by the *William Scoresby* in 1931, a host of researchers have studied this current system (for example: Gunther, 1936; Wooster and Gilmartin, 1961; Wooster and Guillen, 1974; Wyrtki, 1974). The system is composed of four distinct currents, two flowing to the north, two flowing to the south. The Peru Coastal Current, which flows next to shore, and the Peru Oceanic Current (or Humboldt Current), which is situated farther out to sea, flow northward. Between them flows the near-surface Peru Countercurrent. Below the three flows the Peru Undercurrent. The anchoveta tend to congregate in massive shoals in the northern part of the coastal current. The oceanic current, first described in 1803 by Humboldt, the German naturalist, stretches southward from Peru several thousand miles and extends to depths of 700 m. This current was first thought to carry enough cold sub-Antarctic water into the coastal region to explain the low sea surface temperatures observed there. Unfortunately, that theory could not explain why the coldest coastal waters were often found at low latitudes. Only after the emergence of the Ekman theory was there a general recognition of the importance of upwelling to the thermal structure of the coastal waters. Gunther (1936), for example, found that while the appearance of upwelling was somewhat irregular in time and space, it had a very definite relation to the wind, in general agreement with the Ekman theory. Wyrtki (1963), using hydrographic data, deduced that upwelling along the coast is restricted to depths of less than 100 m but that ascending motions at greater depths and further offshore are probably important to the upwelling process. Over large areas, Wyrtki determined vertical motions to be of the order of 10^{-5} to 10^{-4} cm per sec at 100 m. He also concluded that areas of coastal upwelling extend toward the equator and gradually merge with regions of equatorial upwelling.

From time to time (very roughly every seven years), there occurs a major disturbance in the ocean and atmosphere off the west coast of South America. There is some evidence that the atmospheric disturbance may be global in scope (Ramage, 1975). During this disturbance the usual warming of the coastal waters off Peru during Christmastime, originally termed El Niño, continues unabated for much longer than normal. The warming spreads southward during the Peruvian summer and fall until the whole coast of Peru and northern Chile is affected. The results of the warming are catastrophic for the climate, for the ecosystem, and for the Peruvian fishery. El Niño often results in torrential rains in deserts which have had no rain in years; it results in widespread disappearance of the anchovy population; and, it results in the mass mortality of several species of birds dependent on fish for survival. Because of the social, economic, and physical impact of El Niño, a serious effort has been made in recent years to understand and predict it. A representative, but far from comprehensive, list of references on the problem includes: Bjerknes (1961), Paulik (1971), Wyrtki (1973, 1975), Wooster and Guillen (1974), Wyrtki et al. (1976), Hurlburt et al. (1976). We will return to the problem of El Niño in the next section.

North America. The North Pacific high has a major influence on the coastal upwelling off the west coast of North America from Baja California to British Columbia. The equatorward winds are strongest off Baja in April and May, off southern California in May and June, off northern California in June and July, and off Oregon and Washington in July and August (Smith, 1968). The primary current system is the equatorward-flowing California Current (Reid, Roden, and Wyllie, 1958). During fall and early winter a poleward flowing current near the coast, called the Davidson Current, is also observed. Associated with active coastal upwelling is a highly time-dependent equatorward surface jet within 50 km off the coast. This coastal jet has also been observed off northwest and southwest Africa, off Peru, and off the Somalia coast.

Perhaps the most systematically studied coastal upwelling region in the world lies off the Oregon coast. Scientists at the Oregon State University (and more recently other groups participating the Coastal Upwelling Experiments CUE-I, 1972, and CUE-II, 1973) have carefully studied, over more than a decade, the biology, physical oceanography, and marine meteorology of a 100 km² area roughly centered at Newport, Oregon. Mathematical models of the ocean circulation (Thompson, 1974; Peffley and

O'Brien, 1976), the low-level atmospheric circulation (Clancy et al., 1975), and the primary biological productivity (Wroblewski, 1976) have recently been designed to help explain and organize the large amount of data obtained by the field projects. One important result of the studies was the discovery of occasional intense wind-driven upwelling "events" lasting a week or ten days. From this research, pioneering attempts to aid Coho salmon fishermen by predicting areas of potentially good fishing using wind and current information have met with moderate success (Wright et al., 1976).

Northwest Africa. The Canary Current moves equatorward from about 36°N to about 18°N, then joins the Equatorial Current moving southeastward along the Guinea coast toward the equator. There it merges with the northward-flowing Benguela Current. The richest upwelling areas tend to be found off Spanish Sahara and Mauritania. Apparently, upwelling is most intense in spring and summer and migrates northward from winter to summer (Wooster and Reid, 1963). In the summer the upwelling may extend northward to the Straits of Gibraltar. Less intense and less regular upwellings occur in the tropical region along the Gulf of Guinea.

The northwest Africa fishery (and its management) is under-developed but growing rapidly (Crutchfield and Lawson, 1974). The physical oceanography and primary productivity in that region have come under close examination in the past decade, particularly during the multination, multidisciplinary JOINT-I (1974) expedition. A nearshore equatorward jet, a poleward undercurrent and high correlation between currents and surface wind stress appear as persistent features of the reported observations (Mittelstaedt et al., 1975). Despite the similarities between the northwest Africa and Peruvian upwelling areas, the annual fish production off Peru is ten times that off northwest Africa. This difference may be attributed to the single-step food chain and extremely high primary productivity found off Peru, compared to the relatively complex food web and moderate primary productivity found off northwest Africa (Huntsman and Barber, 1976).

Southwest Africa. A region of coastal upwelling extends off the west coast of Africa from about 15°S to 34°S. To the west, in the eastern branch of the South Atlantic anticyclonic gyre, flows the Benguela Current (Hart and Currie, 1960). Smith (1968) identifies the coastal (Benguela) current as analogous to the Peru Coastal Current and the offshore flow as analogous to the Peru Oceanic

Current. Defant (1936) found the region of coldest water to be in the coastal current between 23°S and 31°S. From the charts of Böhnecke (1936), there appears to be most intense upwelling in summer and fall with an equatorward migration of maximum upwelling from the Southern Hemisphere summer to winter.

The upwelling off Namibia appears to have the highest primary production and the largest fish stocks. Until recently the fisheries in this upwelling region were confined to the coastal states of Angola, Namibia (currently called South West Africa), and South Africa, under the jurisdiction of the International Commission for the Southeast Atlantic Fisheries (ICSEAF). During the past decade, distant-water operations from Japan, Spain, the U.S.S.R., and several other foreign nations have become common. Some fisheries experts believe part of this regional fishery has already been fully exploited, particularly the southern hake stocks (Crutchfield and Lawson, 1974). Other species have yet to be fully exploited.

Somalia and Arabia. The upwelling off Somalia and Arabia is unique in that it occurs on the *east* coast of a continent. This is due to the fact that the most vigorous winds blow parallel to the coast from *south* to *north* in late spring and summer. Consistent with the Ekman theory, this situation also produces coastal upwelling. Thus, during the Southwest Monsoon, very intense coastal upwelling occurs along Somalia and Arabia, roughly from the equator to 25°N. Although the sea surface temperatures (SST) in the open oceans of the Northern Hemisphere are warming during this season, SSTs may drop as much as 16°C in the upwelling areas off Muscat-Oman and the Somalia coast. It is believed that these low SSTs influence the atmospheric circulation and may be linked to rainfall distributions over India (Shukla, 1975). The fishery potential for this area is largely underutilized and the region has not been intensively studied in the manner of northwest Africa or Peru, although we learned much from the International Indian Ocean Expedition (1960-65). Recently INDEX (the Indian Ocean Experiment), a part of the Global Atmospheric Research Program, has begun to reexamine the ocean circulation of this upwelling region.

Coastal Upwelling and Terrestrial Deserts

The large-scale atmospheric circulation favorable for biological productivity in the ocean during coastal upwelling also tends to produce the terrestrial coastal deserts evident off South America, and northwest and southwest Africa. The cold surface water produced by the upwelling itself also influences the climate of the adjacent land. Less dramatic evidence of this influence is observed off the west coast of the United States, northwest Africa, and perhaps, on a larger scale, over the western Indian Ocean during the Southwest Monsoon.

The most intense aridity on earth likely occurs along the coasts of Peru and Chile. As Trewartha (1961) notes:

> Because of the general sparseness of weather stations in arid regions, it is almost impossible to indicate with any degree of certainty where the absolutely driest part of the earth is located. Without fear of contradiction, however, it can be stated that the Chilean-Peruvian desert is a strong contender for this honor of being the earth's most arid region over such a large range of latitude. From Piura in northern Peru at about 5 or 6°S to Coquimbo in northern Chile at about 30°S the highest average annual rainfall (1944-1955) at any coastal station is under 50 mm and a number of stations in northern Chile have recorded no rainfall over periods of one or more decades. In the Chilean desert even the drizzle is largely absent, so that here an intensity of aridity prevails that, so far as is known, is not equaled in any other desert of the earth.

Certainly the primary physical mechanism responsible for this coastal desert is the strong subsidence associated with the eastern flank of the South Pacific anticyclone. This feature is highly permanent, owing to the impeding and diverting atmospheric zonal flow patterns and migrating weather systems of the South American Andes. However, this mechanism alone does not appear sufficient to reduce rainfall to the minimal levels observed. There is evidence that the most intense aridity is concentrated in a narrow strip along the coast and that rainfall increases both seaward and inland. Where points of land project westward into the ocean, rainfall is lowest. Cold upwelled waters (as much as 8°C colder than water 100 km seaward), occurring within 50 km of the coast of Peru and Chile, stabilize the lower marine boundary layer, intensify and lower the temperature inversion that exists there, and enhance the development of fog, low stratus, and sometimes drizzle along the coast. The cold water and the contrast of temperature and frictional drag between land and sea may also contribute to the strength of the subsiding air over this coastal desert. Unquestionably, active coastal upwelling is an important factor in the creation of the Peru-Chile coastal desert, perhaps the driest desert on earth.

Along the coast of southwest Africa lies another desert adjacent to a coastal upwelling zone. The Namib has a climate characterized by intense aridity, negative temperature anomalies, small annual and diurnal ranges of temperature, and high frequency of fog, low stratus, and drizzle. The Namib has an extensive latitudinal range of about 15° along a narrow coastal strip extending only a few tens of kilometers inland. Several coastal stations average less than 25 mm of annual rainfall (Trewartha, 1961), with the most intense aridity concentrated around 22° to 28°S. As along the Peru-Chile coast, rainfall amounts increase rapidly inland. Unlike South America, the southwest African coastline does not have a pronounced westward bend toward the equator. Therefore, the coast does not maintain contact with the cold coastal current nearly as far north as the Peruvian coast maintains contact with the Peru Coastal Current. Further, no coastal cordillera exists to influence the circulation around the eastern edge of the oceanic high. While Luanda at 9°S has 339 mm of rainfall yearly, at the same latitude along the Peruvian coast annual rainfall is less than 50 mm.

The coastal upwelling region off northwest Africa also adjoins a dry western littoral. However, there exists at these latitudes a much larger-scale belt of subsiding air, part of the climate regime extending at these latitudes across the Atlantic, to North Africa, Arabia, western Asia, and northwestern Pakistan. The western reaches of the Sahara merge with the superarid coastal zone of northwest Africa. The cool Canary Current and the intense coastal upwelling influence the coastal climate southward at least to 15°N. At Port Etienne at 21°N, for example, rainfall is only about 28 mm per year. The coastline bends eastward south of Cap Blanc and, hence, the dry coastal zone does not extend to such low latitudes as it does off Peru and southwest Africa.

The major climatic influence along the west coast of the United States is the North Pacific subtropical high. Since the North American coastline is situated at considerably higher latitudes than the other coastal upwelling regions we have examined, the influence of this high is most important during summer when it extends farthest poleward. Even during summer there is little evidence of vigorous coastal upwelling or strong atmospheric subsidence poleward of 50°N (Lydolph, 1955). However, to the south, along the coast, cool surface temperatures and a precipitation minimum are commonly observed in summer. During winter, when the Pacific high migrates southward, major storms provide the coastal region with significant rainfall and coastal upwelling is not observed. As in the other major coastal upwelling regions, the primary mechanisms suppressing atmospheric convection are large-scale subsidence from the oceanic high and the stabilizing influence of the cold upwelled waters. Further south, under the strengthening influence of the subtropical high, precipitation amounts decrease. From northern California southward to Baja, California, the coastal climate ranges from cool summer marine to mediterranean to steppe to subtropical desert.

The coldest coastal waters during summer occur near 40°N. For example, during August 1973, sea surface temperatures as low as 7° C were observed within 20 km of the Oregon coast (Holladay and O'Brien, 1975). Occasionally the cool, moist marine air along the coast is advected inland to the warm, dry interior valleys by intense sea breeze circulations (Johnson and O'Brien, 1973). Perhaps the most well-known climatic influence of these upwelled cold waters is in San Francisco in northern California. That city boasts one of the coldest mean summertime temperatures in the continental United States. The July mean temperature is 59°F and the daily summertime maximum is 65°F. Precipitation is also at a minimum during this active upwelling season. The cool upwelled waters and the high incidence of stratus clouds which reduce insolation are responsible for this remarkable climatic anomaly.

Ocean Desertification

The reduction of biological diversity and abundance in the sea is a problem as serious, though not as well known or well studied, as that encountered on land. While the land may suffer from overgrazing, the sea may suffer from overfishing. While the land may experience long-term weather changes that affect crop yields, the sea may experience changes in currents and thermal structure which produce severe dislocations in fisheries; while the land may be appropriated for many uses other than food production, the sea and seabed, particularly along coastal margins, may be utilized as a dumping ground, a mining pit, and in many other ways incompatible with the ocean ecosystem. While terrestrial productivity may suffer degradation from pollutants of all types, the sea and the seabed become the ultimate sink for most of them.

The potential loss of biological productivity in the sea gives cause for concern and reason for study. The concept of presently productive ocean regions losing their fertility and becoming new ocean deserts due to anthropogenic causes is not only fascinating and rather unusual, it is also very real.

Natural Variability

In some instances, plants and animal species in the sea undergo natural fluctuations in their population and distribution which are clearly related to observable changes in the physical environment. The decrease in anchovy catch and bird population during El Niño is a dramatic example of such a natural fluctuation. Major oceanographic changes are quite noticeable during the El Niño, with warmer than normal sea surface temperatures and lower surface salinities. Other natural fluctuations, such as the disastrous blooms of the "red tide" and the disappearance of large populations of the California sardine are not so obviously related to changes in the physical environment. These natural fluctuations, and our enormous ignorance of ecosystem dynamics, make it extremely difficult to attribute reductions in ocean productivity to a particular "unnatural," anthropogenic cause. Awareness of this natural variability is crucial when attempting to interpret data which, at first glance, might suggest decreases in a species population due to human influence.

There occurs an analogous situation in the problem of terrestrial desertification. Both natural climatic variability and anthropogenic causes have been suggested as contributing to extended droughts in the Sahel. MacLeod (1976), Charney (1974), and others have proposed that overgrazing, erosion, deforestation, and generally poor land management could lead to increased desertification there. Namias (1974), Winstanley (1974), and others have theorized that the changes in large-scale components of the general circulation played an important role in creating the Sahelian drought. Unfortunately, our data and our theoretical knowledge do not permit us to decide the relative influence of each factor. It seems likely that both natural variability and man's activities played a significant role—with overgrazing and other related activities aggravating the problems initiated by climate fluctuations.

In the case of the oceans, we are faced with the similar task of assessing the relative importance of natural and anthropogenic influences on the loss of utilizable biological productivity. Such a task appears enormously difficult and the results seem certain to be subjected to considerable debate. The problem has been succinctly stated by Dickie (1975):

> The primary difficulty faced by the biological oceanographer in analyzing changes in fisheries is that the effects that arise from both economic and environmental causes look very nearly the same as those that would result if excessive fishing were damaging the stock productivity. Given a world situation in which economic difficulties are almost the rule rather than the exception, and where there are widespread predictions that climate is at the beginning of a long downward trend in temperatures, to say nothing of the dangers of pollution, must it be concluded from the present decline in catches that the world high seas fisheries have reached a practical upper limit to the sustained yield? An answer to this question becomes especially urgent in the present situation where more and more of the world's fisheries are subject to regulations on economic grounds, but often with the claim that restrictions are necessary in the interest of conservation, at a time when there is increased world need for protein foods.

There is considerable evidence that natural changes in the ocean's climate have adversely affected fish stocks. Dickson and Lamb (1971) identified changes in the environmental conditions near Iceland as affecting the catch of Icelandic herring. From the mid-1960s to 1970 the herring catch dropped from 750,000 tn to 50,000 tn. During the same period the ice cover in the region increased, influencing the migration routes of the herring and perhaps keeping the fish out of range of the Icelandic fleet. Poor harvests of Alaskan salmon in 1973 and 1974 have been partially attributed to cold water anomalies near the Aleutian Islands (Johnson, 1976). Nelsen et al. (1975) found a relationship between the wide range in year-class size of Atlantic menhaden and an index of surface water drift. Parrish (1976) related variations in surface water drift in the spawning grounds of the Pacific mackerel off Baja California to variations in the year-class size.

Perhaps the best case history of a natural fluctuation in climate which has resulted in reduced fish yields is the now famous story of El Niño. Yet even in this case we know very little about how that reduction in harvestable fish is accomplished. Further, there is

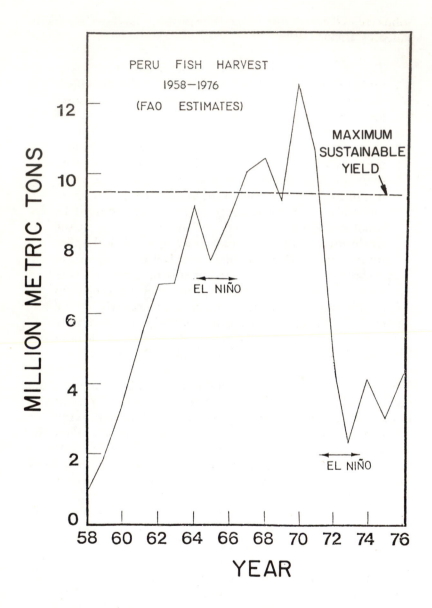

Figure 6.2. Peru fish harvest, 1958-76. The 1976 figure is estimated from early trends (FAO and U.S. Department of Agriculture data). Dashed line represents best estimate of maximum sustainable yield (Paulik, 1971).

some evidence to suggest that during the 1972 El Niño, man's fishing activity and natural environmental changes acted in concert to seriously threaten the future of the Peruvian anchovy fishery.

Until World War II the bird guano industry of Peru greatly overshadowed the fishery. Since the guano birds feed on fish, the two industries were hardly compatible. In 1950 a small fish meal factory was quietly constructed along the Peruvian coast and, before the guano industry could react, an incredible fishing boom seized the Peruvian economy. (For an excellent history of this see Paulik, 1971.) By 1963, Peru, with a population of 13 million, had become the leading fishing nation in the world, harvesting 15% of the world's total catch. As the fishery grew the bird population shrank. In 1957 about 28 million guano birds were estimated; by 1970 the estimate had decreased to about 2 million. The population has not significantly increased in this decade (Kestevan, 1976).

As the fish harvest soared in the early 1960s the United Nations FAO and the Peruvian government, under the Instituto del Mar del Peru (IMARPE), began working together to study, monitor, and regulate the fishery. Increased fishing effort and catch capacity was controlled by regulation through closed seasons and catch quotas. Since quotas were rapidly taken by competing fish companies, the days at sea each year were reduced. However, these days were continually more concentrated in periods when new recruits joined the fish stocks, hence fishing pressure on fish in their first year increased (Valdivia, 1974).

In 1970, under FAO and IMARPE sponsorship, a panel of fishery experts recommended that the maximum sustainable yield for the Peru Current anchoveta stock was about 9.5 mmt. However, by 1970, the harvesting and processing capabilities of the Peruvian fishery were far above the capacity required to efficiently land 9.5 mmt. Although 9.5 mmt was harvested for the 1969-70 season by 28 April 1970, an additional ten days of fishing yielded an incredible 1 mmt. Before the season was closed another 0.5 mmt was caught. The total catch for calendar year 1970: 12.6 mmt.

Although a mild El Niño had forced a brief pause in catch growth during 1965-66, the fishing industry was hardly prepared for the events of 1972 (figure 6.2). As early as March 1971, surface waters in the equatorial Pacific east of 110°W became significantly warmer than in the previous year (Wooster and Guillen, 1974). On the northern Peruvian coast the warming first became evident in

Table 6.2.   Percentage of Peruvian Fish Harvest Taken at
             Various Distances Offshore During March,
             1970–1974 (from Valdivia, 1974)

Distance from Coast (miles)	1970	1971	1972	1973
0 – 10	42	47	91	88
10 – 20	33	33	7	11
20 – 30	16	13	1	1
30 – 40	5	5	1	–
40 – 50	4	1	–	–
50 – 60	–	1	–	–

February 1972, when low salinity surface water was found as far south as 10°S.

The catch statistics taken in the latter half of 1971 indicated anchovy concentrations close to shore (within 20 mi). In March 1972, when the fishing season reopened, these concentrations remained near shore, but shrank latitudinally, captures occurring only south of 10°S. During March, with El Niño already evident, a record catch rate (over 170,000 tn per day) was realized within a few miles of the coast! As shown in table 6.2 (from Valdivia, 1974) the percentage of the total catch taken within 10 mi of the coast during March doubled in 1972, as compared to 1970 and 1971. By April the schools had retreated south of 12°S, then to 14°S, and by June operations became uneconomical and were halted.

The next year yielded a total catch of only 2.3 mmt, less than 20% of the 1970 harvest. The 1974 and 1975 harvests were also extremely poor. The estimated anchovy catch for 1976 ranges between 3.5 and 4 mmt. Low exports of fish meal and no exports of fish oil (due to low oil yield as well as the small size of the catch) were anticipated for 1976.

It seems clear that natural variability of the ocean-atmosphere system was responsible for the decline in the Peruvian anchovy harvest during 1972-73, as it has been in the past. However, two factors appear to have acted in catastrophic harmony in 1972 to produce a severe, perhaps permanent, dislocation in the anchovy fishery. First, during El Niño, natural oceanographic conditions changed in such a manner as to increase the apparent concentration of anchovy very near the coast. Second, at the time of El Niño, the fishing fleet had grown far too large for controlled collection of fish under conditions of such high concentrations. While we do not know why the fishery has failed to make a rapid recovery, we do know that during El Niño the fish driven close to shore were scooped up in record numbers just prior to the decline and collapse of the fishery. There is some reason to believe that man and nature cooperated in decimating a phenomenal fishery which may, or may not, recover.

Pollution

Pollution of lakes, rivers, estuaries, coastal margins, and the open ocean has received considerable popular and scientific attention during the past decade. Yet, as Hela (1971) states, even the scientific literature tends to be biased: "Every pessimist, regardless of the reliability of his scientific results, is anxious to publish his warnings, while every optimist, even following a serious scientific study of the situation, is rather reluctant to present his sensational finding to the public at large." Whatever their personal views, however, most scientists would likely agree that when discussing loss of biological productivity in the sea due to pollution, the key word is ignorance. We know very little about the *natural* state of the oceans and its variability, much less the effects of man-made pollutants on that state. In these brief pages a comprehensive, balanced review of marine pollution is impossible. A more realistic goal is to offer an enumeration of well-known sources of marine pollution and present documented evidence for loss of marine biological productivity.

It is an unfortunate fact that the most productive regions of the ocean, the coastal zones, are also those nearest human activity and most threatened by pollution. Not surprisingly, most evidence for decreased biological productivity from pollution has come from studies of these zones. The sensationalized evidence—oil spills, thermal pollution from power plants, and destruction of marshlands—is highly localized, easily observed, and readily reported. However, it is the slow, insidious, and undramatic degradation of the marine environment that is likely to be of most concern in the biological desertification of the sea.

The substances produced by man which eventually find their way into the ocean number in the tens of thousands. Most exist in such small quantities or are so readily destroyed that they pose no long-term threat to marine productivity. Some occur naturally and fall into the "background" levels already present. A few substances have been identified as having *potentially* long-term effects on the marine environment due to attributes of persistence, quantity, and toxicity. A partial list of present serious pollutants includes crude oil and distilled by-products, halogenated hydrocarbons (DDT and PCB in particular), inorganic chemicals (including heavy metals), domestic wastes, plastics, transuranic elements, and warm-water discharges from power plants. It is likely that there are others not yet identified as harmful, either due to difficulty in observing them or to their recent creation. The reader is referred to several excellent reviews of marine pollution for discussions of each of these substances (see Hood, 1971; Hela, 1971; *Oceanus*, fall 1974).

It is often simpler to identify oceanic pollutants than it is to demonstrate their contribution to ocean desertification. Even if organisms themselves are not seriously affected by a particular pollutant, they may concentrate that pollutant in tissues to such an extent that they are no longer safe for human consumption. This "economic" desertification of the sea is no less a loss of a potential food source.

Examples of short-term destruction of species of fish, shellfish, and aquatic birds due to marine pollution are numerous. Oil spill effects have been particularly visible (Blumer, 1970). Accumulation of chlorinated hydrocarbons have caused reproduction failures in sea birds (Risebrough, 1971). DDT effects upon fish reproduction have been noted by Burdick et al. (1964) and Macek (1968). In 1969 jack mackerel (*Trachurus symmetricus*) from the Pacific were confiscated and condemned by the U.S. Food and Drug Administration for high—10 parts per million (ppm)—concentrations of DDT compounds. PCBs have been found in wildlife and fish in Holland, off the Scottish coast, off Germany, in the Baltic Sea, and in Japan during the past decade (Jackson, 1976). In 1976 New York State closed the Hudson River to all commercial fishing except for sturgeon, shad, and goldfish. The EPA reported concentrations of PCB as high as 350 ppm in the Hudson river fish. The FDA maximum is 5 ppm. The Canadian maximum is 2 ppm.

One hopeful sign is that concentrations of DDT and PCBs in the open ocean appear to be decreasing due to recent restrictions on their use. Harvey (1974) reports that PCB concentrations in North Atlantic waters have decreased fortyfold since 1972, although a constant influx is still being detected. DDT levels have dropped sharply in coastal waters of North America. Unfortunately, while regulations are imposed on these compounds, new chemicals whose long-term effects are unknown are produced. The recent Kepone pesticide contamination of Virginia's James River and the Chesapeake Bay is yet another example of the difficulties involved in detecting and assessing the effects of pollutants on marine ecosystems.

The most widespread cause of coastal zone pollution in the United States is municipal wastes. More than 60% of the population lives within 250 mi of the coast. About 29 billion gallons of untreated sewage are dumped into the coast waters every *day* (Smith, 1974). Halstead (1970) has observed that:

> Thousands of halibut, croaker, sea-bass, sole, sand-dabs, and other shore fishes, in the vicinity of sewage outfalls, have had an alarmingly high incidence of cancerous growths, skin ulcers, malformations, emaciations, and genetic changes. These pathological disturbances are believed to be due to the toxic effects of pollutants. The precise causative agents are unknown. The possible public health implications to man are of growing concern.

Economic losses, presumably related to marine productivity losses, demonstrate the extent to which coastal zones have been affected by pollution. In 1970, the National Marine Fisheries Service estimated losses from polluted oysters and clams at $12 million. Including shrimp, lobster, and crabs, the Council on Environmental Quality estimated the loss at $63 million. Smith (1974) has identified several specific economic losses due to pollution, sedimentation, and dredging of shellfish beds and fishing waters:

> — In Connecticut, a combination of pollution and marsh destruction has reduced the annual harvest of clams from $20 million during the 1920s (equivalent to $48 million at today's prices) to only $1.5 million during the 1970s.
> — In Galveston Bay the catch of shrimp declined, even with increased efforts at harvesting, from 14.2 million pounds in 1962 to 1.9 million pounds in 1966, as indus-

trial, domestic, and oil pollution increased.

— In Raritan Bay, between New Jersey and New York, the current harvest of herd clams is worth $40,000 a year. With clean water, the annual harvest could be $3.85 million. The present annual finfish harvest, worth $200,000, could be doubled if the waters were clean.

— In the Chesapeake Bay, over 50 percent of the upper estuarine areas for fish spawning and shellfishing were destroyed between 1800 and 1950 by dredging, filling and pollution.

Many other localized examples of the effects of pollution on marine productivity could be cited. However, the point is rather clear: gradual pollution of the coastal zone is a major contribution to ocean desertification. The vision of an ocean desert, however temporary, localized, and "unscientific," was forcefully illustrated in a recent (3 August 1976) letter appearing in the Washington, D.C., *Star*:

I am writing this letter in an attempt to bring to public attention a problem about which I feel most people are unaware.

I was scuba diving off the coast of New Jersey the weekend before last about 70 miles south of New York City, ranging from two to 16 miles off the coast. During the two days we visited eight sites. The scene on six of them was one of total and complete destruction of sea life on the ocean floor.

There was simply nothing alive. Dead fish were lying everywhere in varying states of decomposition, as were lobsters, crabs and clams. Even the anemones which cover the shipwrecks were dead. Visibility was so bad that one could not even see his own fins at times. One of the members of my party brought up about six dead lobsters with the intention of trying to get them analyzed at a laboratory. They smelled strongly of raw sewage.

I have heard several theories as to the reason for this particular kill. Whatever the particular combination of unusual winds or currents that has made this year so much worse than most, it is apparent that man has been too careless in the ocean dumping of wastes.

I don't know the exact extent of this destruction but I do

know that it extends north of New York City and discussions with other divers have revealed that it extends south of Rehoboth, and the area is spreading.

I am sure that if a similar catastrophe occurred on land, the public reaction would prompt some action to remedy it.

Overfishing

Traditional fisheries are presently in difficulty on a global scale. Popular fish species which are easily marketed—herring, cod, haddock, flounder, sardines—are now persistently fished above their level of maximum sustainable yield (Edwards and Hennemuth, 1975). As an example, the International Convention for the Northwest Atlantic Fisheries estimated that in the early 1970s the effort applied annually to the groundfish of Georges Bank exceeded by more than 30% that required to take the maximum sustainable yield (Storer and Bockstael, 1975). Development of distant water fleets of large fishing and support vessels and higher levels of fishing effort have resulted in overfishing, despite admirable efforts of numerous regulatory commissions. Yet the total catch of the world fisheries, including aquatic plants, animals, and marine fish, *declined* in the early part of this decade. In contrast, the total catch doubled from 1950 to 1960 and increased half again between 1960 and 1970 (FAO, 1974).

Pressures for additional fishing efforts appear to be building. Presently, developing countries contain half the world's population but consume only about one-fourth the world's fish supplies, though fish is a traditional part of the diet in many developing countries. The demand for additional fish products is likely to grow in response to rising populations, higher costs for agricultural products due to rising fertilizer and energy costs, and the progressive loss of arable lands. It is inconceivable that efforts to extract additional protein from the sea will remain at present levels or decrease for the remainder of this century.

By itself, overfishing does not normally pose the serious threat of extinction of a particular species. Instead, overfishing leads to conditions in which species of marketable size become so scarce that costs of production equal or exceed earnings (Eddie and Insull, 1973). The danger is that when a species becomes overfished the biological niche it occupies may be overtaken by another species less economically harvestable or marketable. The joint effects of overfishing and natural variability of the population may reduce the ability of the species to regain that niche following a decline (Dickie, 1973). The net result of this "economic desertification" is nevertheless a loss of biological productivity useful to man and, in the context of our discussion, represents a form of ocean desertification.

A classic example of the impact of natural variability and overfishing on a fish species is the case of the California sardine (*Sardinops caerula*). The fact that fishery scientists are still debating the relative impact of these two factors on the decline of the fishery is a testament to our lack of knowledge concerning population dynamics in marine species. From 1916 to 1936 the annual catch of California sardines rose from 28,000 to nearly 800,000 tn (Murphy, 1966). During the following ten years the annual catch fluctuated between 500,000 and 700,000 tn. In 1945 it dropped below 200,000 tn, rose slightly for several years, then collapsed in 1952 to about 20,000 tn—where it remains.

Murphy suggests that there is little likelihood that the sardine population would have declined in the absence of fishing pressure. The biological niche vacated by the California sardine was filled by the northern anchovy and the sardine population has remained low (Joyner, 1971). Dickie (1973) argues that sediment records indicate that the sardine has been replaced by the anchovy in earlier times and that overfishing may have only increased the probability that this replacement would occur again.

Overfishing has become particularly evident during the past decade. Moorcraft (1973) has noted that in 1949 a United Nations conference on conservation and utilization of resources identified only a few species, such as plaice, halibut, and salmon, as being overexploited. The conference identified 30 underexploited stocks. In 1969 half of those were identified by a similar conference as being overexploited or nearly so. The evolution of the Peruvian anchovy fishery demonstrates how rapidly the potential for overfishing can be generated.

As fishing operations have grown, so too have the number of international fisheries commissions, whose primary functions are to protect fishery resources and reduce international conflicts without assuming or designating exclusive resource rights (Storer and Bockstael, 1975). Unilateral claims to extensive fishery jurisdictions have also increased. Yet these restrictions thus far have not been sufficient to eliminate overfishing. As noted by the National Academy of Sciences Committee on Oceanography (1967):

> Ocean fishing power on a world-wide basis is growing at a much more rapid rate than the means of measuring its effect on the fish stocks it is being applied against. . . . This whole field of marine science is being swamped by the developing fishing power. The nations devote their ocean research funds to the development of fisheries, but they are laggardly in providing research funds for the detailed biological and population dynamics research which alone can give guidance in the solution of the problems which expanding fishing creates. Nations do not like to put their fishermen under regulation even to provide the conservation that they have agreed to provide unless the scientific needs for the regulation are established and they do not like to put up money to provide the research needed either to determine the need for regulation or the form it should take.

The Third United Nations Conference on the Law of the Sea (UNCLOS) failed to agree upon a comprehensive, detailed, and widely accepted ocean treaty, a portion of which would address the problem of overfishing. Under strong domestic and international pressure, however, the United States agreed in the Caracas UN-CLOS session of 1974 to accept a 200-mi resource zone. The draft articles have been summarized by Osgood et al. (1976):

> The articles stipulated that the coastal state would exercise jurisdiction and sovereign and exclusive rights for the purpose of exploring and exploiting the natural resources of the 200-mile zone. In regulating fisheries within the zone, the coastal state would insure conservation and full utilization of the resources. The coastal state would establish the allowable catch, within which limit it would harvest up to its full capacity.

> Traditional fishing states and states of the region would be licensed for a reasonable fee to harvest the remainder of the allowable catch. Fishing for anadromous species

would be prohibited except as authorized by the state of origin. And management of highly migratory species would be governed by regulations established by regional or international organizations. The organization would establish allowable catch, allocation regulations and rules for the collection and payment of licensing fees.

Recently the United States unilaterally passed legislation establishing the 200-mi resource zone off its shores. Other coastal states have adopted similar or more restrictive jurisdictional policies. The trend toward coastal state control of the resource zone should ultimately reduce the number of distant-water factory ships. Domestic fishing industries should grow. The seas should become less a kind of "commons" and more an extension of the economic interest of the coastal state. As Storer and Bockstael (1975) have observed, ". . . the advent of extended jurisdictions within the content of an overall law of the sea treaty presents the opportunity for conservation and efficient production, and, as such, provides the greatest hope of achieving the optimum utilization of the sea's living resources."

Concluding Remarks

The finite limits to the earth's resources and carrying capacity have only recently been fully appreciated (SCEP, 1970). The concept of a limitless sea is perhaps the most difficult aspect of the "infinity-mythology" to dispel. We are now beginning to realize that the sea is not a homogeneous, steady-state entity which quickly renders harmless our wastes while maintaining bounteous food resources in the face of enormous fishing pressures. The sea is neither inexhaustible nor can it long tolerate unbridled exploitation.

A necessary first step in addressing the problem of overexploitation of the sea is enlightenment. This chapter has set forth the hypothesis that the sea, as the land, is susceptible to extreme losses of biological productivity due to natural and anthropogenic factors. It has emphasized that most of the sea is a biological desert and that the areas of high productivity are highly concentrated in coastal zones closest to man's pollution. Since coastal upwelling zones account for roughly half the world's fish harvest while comprising only 0.1% of the ocean surface area, they have come under particular scrutiny. Since the climatic conditions favorable for above average ocean productivity are often conducive to the formation of coastal deserts, these areas have been closely examined. In effect, nature has compensated for the barren land with the abundant sea in these regions. Unfortunately, few local residents receive that compensation.

Natural variability of the ocean's climate has been identified as one primary cause for reduction of productivity. This form of ocean desertification is particularly difficult to separate from the impact of overfishing. Overfishing itself may reduce the ability of a fish species to recover from natural fluctuations in population. The El Niño occurrences in the Peruvian upwelling region and the decline of the California sardine fishery are examples of the influence of natural variability and overfishing on productive fisheries.

Pollution has also contributed to ocean desertification, as well as desertification of inland waters. Coastal urbanization, the filling of marshlands, and the plodding development of adequate waste disposal methods have already threatened some coastal areas with serious loss of biological and economic productivity. Our ignorance of the effects of newly created pollutants and their time scales of activity is a significant obstacle to the prevention of further loss of marine productivity. Of particular concern is the problem of nondegradable or slowly degradable materials: plastics, hydrocarbons, radioactive isotopes, heavy metals, and the like. While the present problems of desertification appear to be local, we have no assurance that they will remain so under the prospects of increased population and technological growth.

As with the land, the sea can be made more productive. One approach is through fish or plant farming—mariculture. However, significant contributions to the world food supply are not likely to come through energy-intensive farming of the sea. Rather, they must come from indigenous populations who have a nutritional need and available unpolluted coastal and inland waters and who employ labor-intensive methods. Ryther (1975) estimates that 100 million acres or more of coastal wetlands have the potential for utilization. Some ocean deserts may be "reclaimed," as have been some inland lakes, by reductions in pollution. Overfishing practices also must be abolished. Unconventional but underutilized marine species can be harvested if processing and marketing techniques can be made economically viable (Rathjen, 1975). At present the problem of ocean desertification appears to be largely localized, often man-made, and usually reversible. Therein may lie the greatest hope for its cessation.

Acknowledgments

Dr. Michael Glantz provided the original inspiration for this contribution. His assistance was essential for its completion. George Wooten provided helpful suggestions concerning coastal deserts. Dr. Joseph Wroblewski and M. B. Peffley offered numerous constructive technical comments and corrections. I thank Margaret Mikota for typing many drafts and for providing exceptional assistance. I appreciate the work of my wife, Brenda, who reviewed the drafts and offered helpful criticisms. This work was completed while the author was under subcontract to the Naval Research Laboratory, Washington, D.C., and participating in the NORPAX and INDEX programs of ONR and IDOE.

Note

1. This productivity is generally higher in the eastern equatorial ocean than in the western ocean since the thermocline, which effectively separates the warm upper ocean from the cool deep water, deepens toward the west.

References

Allen, J. S., 1976. Some aspects of the forced wave response of stratified coastal regions. *J. Phys. Oceanogr., 6*: 113-119.

Bakun, A., 1973. Coastal upwelling indices, west coast of North America, 1946-1971. *Tech. Rep. NMFS SSRF 671*, NOAA, Seattle, Wash., 103 pp.

Bjerknes, J., 1961. El Niño study based on analysis of surface temperatures, 1935-1957. *Bull. Inter. Trop. Tuna Comm., 8*: 33-165.

Blumer, M., 1970. Oil contamination and the living resources of the sea. FAO seminar on methods, measurement and monitoring of pollutants in the marine environment. Rome, 4-10 December, 1970; final report, 94 pp.

Böhnecke, G., 1936. Temperatur, salzgehalt und Dichte an der Oberfläches des Atlantischen Ozeans. *Atlas. "Meteor" Rep., 5,* 74 charts.

Burdick, G. E., E. J. Harris, H. J. Dean, T. M. Walker, J. Shea, and D. Colby, 1964. The accumulation of DDT in lake trout and the effect on reproduction. *Trans. Amer. Fish. Soc., 93*: 127-136.

Charney, J. G., 1974. Dynamics of desert and drought in the Sahel. *Symons Lecture to the Royal Meteorological Society,* 20 March.

Clancy, R. M., H. E. Hurlburt, J. D. Thompson, and J. D. Lee, 1975. The development of a numerical, mesoscale, air-sea interaction model. *NCAR Technical Note Proc-107,* 1-30.

Crutchfield, J. A., and R. Lawson, 1974. West African marine fisheries: Alternatives for management. Resources for the Future, Inc., Washington, D.C., 64 pp.

Cushing, D. H., 1969. Upwelling and fish production. *FAO Fish. Tech. Paper, 84,* 38 pp.

Defant, A., 1936. Das Kaltwasserauftriefsgebiet vor der Küste Südwest-afrikas. *Landerkdl. Forsch., Festschr. N. Krebs,* pp. 52-66.

Defant, A., 1961. *Physical Oceanography,* Vol. 1, Elmsford, New York: Pergamon Press, 729 pp.

Dickie, L. M., 1973. Interaction between fishery management and environmental protection. *J. Fish. Res. Board Can., 30,* Pt. 2, pp. 2496-2505.

Dickie, L. M., 1975. Problems in prediction. *Oceanus, 18*: 30-35.

Dickson, R. R., and H. H. Lamb, 1971. A review of recent hydrometeorological events in the North Atlantic sector. Contribution 1. International Commission for the Northwest Atlantic Fisheries Environmental Symposium, 18 and 19 May, Dartmouth, Nova Scotia. Bedford Institute of Oceanography.

Eddie, G. C., and A. D. Insull, 1973. Impact of technical development on the problems and opportunities in world fisheries and their management. *J. Fish. Res. Board Can., 30,* Pt. 2: 2490-2495.

Edwards, R., and R. Hennemuth, 1975. Maximum yield: assessment and attainment. *Oceanus, 18*: 3-9.

Ekman, V. W., 1905. On the influence of the earth's rotation on ocean currents. *Arkiv. Mat. Astron. Fysik, 12*: 1-52.

FAO, 1974. *Yearbook of Fishery Statistics,* Vol. 39.

Gulland, J. A., 1968. Population dynamics of the Peruvian anchoveta. *FAO Fish. Tech. Paper, 72*, 29 pp.

Gunther, E. R., 1936. A report on oceanographical investigations in the Peru Coastal Current. *Discovery Rep., 13*: 107-276.

Halstead, B. W., 1970. Toxicity of marine organisms caused by pollutants. FAO Seminar on methods of detection, measurement, and monitoring of pollutants in the marine environment. Rome, 4-10 December, 1970; final report, 94 pp.

Hart, T. J., and R. I. Currie, 1960. The Benuela Current. *Discovery Rep., 31*: 123-298.

Harvey, G. R., 1974. DDT and PCB in the Atlantic. *Oceanus, 18*: 18-23.

Hela, I., 1971. Marine productivity and pollution. In *The Environmental Future*, Proceedings of the first International Conference on Environmental Future, Nicholas Polunin, ed. New York: Barnes & Noble Books. Pp. 249-272.

Holladay, C. G., and J. J. O'Brien, 1975. Mesoscale variability of sea surface temperatures. *J. Phys. Oceanogr., 6*: 761-772.

Hood, D. W., and C. P. McRoy, 1971. Uses of the ocean. In *Impingement of Man on the Oceans*, D. W. Hood, ed. New York: John Wiley & Sons. Pp. 667-698.

Houghton, R. W., and T. Beer, 1976. Wave propagation during the Ghana upwelling. *J. Geophys. Res., 81*: 4423-4429.

Huntsman, S. A., and R. T. Barber, 1976. Primary production in the upwelling region off Northwest Africa—A comparison with Peru. *CUEA Newsletter*, 5 (1): 2-5.

Hurlburt, H. E., and J. D. Thompson, 1976. A numerical model of the Somali Current. *J. Phys. Oceanogr., 6*: 646-664.

Hurlburt, H. E., J. C. Kindle, and J. J. O'Brien, 1976. A numerical simulation of the onset of El Niño. *J. Phys. Oceanogr., 6*: 621-631.

Jackson, T. G., 1976. PCB time bomb. *Oceans, 9* (4): 58-63.

Johnson, A., and J. J. O'Brien, 1973. A study of an Oregon sea breeze event. *J. Appl. Meteor., 12*: 1267-1283.

Johnson, J. H., 1976. Effects of climate change on marine food production. In *Climate and Food—Climatic Fluctuation and U.S. Agricultural Production*. National Academy of Sciences, Washington, D.C., 212 pp.

Joyner, T., 1971. Resource exploitation-living. In *The Impingement of Man on the Oceans*. D. W. Hood, ed. New York: John Wiley & Sons. Pp. 529-551.

Kestevan, G. L., 1976. Recovery of the anchovy and "El Niño." *CUEA Newsletter*, 5 (3): 17-22.

Lydolph, P. E., 1955. A comparative analysis of the dry western littorals. Ph.D. thesis, University of Wisconsin, 59 pp.

Macek, K. J., 1968. Growth and resistance to stress in brook trout fed sublethal levels of DDT. *J. Fish. Res. Board Can., 25*: 2443-2451.

MacLeod, N. H., 1976. Dust in the Sahel: Cause of drought? In *The Politics of Natural Disaster*, M. H. Glantz, ed. New York: Praeger Publishers. Pp. 214-231.

Mittelstaedt, E., D. Pillsbury, and R. L. Smith, 1975. Flow patterns in the Northwest African upwelling area. *Deutsch. Hydrogr. Zeit., 28*: 145-167.

Moorcraft, C., 1973. *Must the seas die?* Gambit Publishers, Boston, 201 pp.

Moore, D. W., and S. G. H. Philander, 1976. Modelling the equatorial oceanic circulation. In *The Sea, Vol. 6* (in press).

Murphy, G. I., 1966. Population biology of the Pacific sardine (*Sardinops caerula*). *Proc. Cal. Acad. Sci.*, 4th Ser. 34: 1.

Namias, J., 1974. Suggestions for research leading to long-range precipitation forecasting for the tropics. Paper presented to the *International Tropical Meteorology Meeting*, 31 January–7 February, Nairobi, Kenya. Sponsored by the American Meteorological Society.

National Academy of Sciences, 1975. *Productivity of World Ecosystems*. Proceedings of a symposium, Washington, D.C., 166 pp.

Nelsen, L. F., W. C. Burrows, and F. C. Stickler, 1975. Recognizing productive, energy-efficient agriculture in the complex U.S. food system. ASAE Paper 75-7505. St. Joseph, Michigan, American Society of Agricultural Engineers.

Nielsen, E. S., and E. A. Jensen, 1957. *Galathea Report*, F. Bruun et al., eds. London: Allen and Unwin. 1: 49.

Oceanus, 1974. Marine Pollution. Vol. 18 (1), 65 pp.

Osgood, R. E., A. L. Hollick, C. S. Pearson, and J. C. Orr, 1976. *Toward a National Ocean Policy: 1976 and Beyond*. Ocean Policy Project, The Johns Hopkins University, School of Advanced International Studies, Washington, D.C., 207 pp.

Parrish, R. H., 1976. Environmental-dependent recruitment models and exploitation simulations of the California Current stock of Pacific Mackerel (Scomber japonicus). Ph.D. diss. Oregon State University, Corvallis.

Paulik, G. J., 1971. Anchovies, birds, and fishermen in the Peru Current. In *Environment, Resources, Pollution, and Society*,

W. W. Murdock, ed. Sinuala Associates, Inc. Pp. 156-185.

Peffley, J. B., and J. J. O'Brien, 1976. A three-dimensional simulation of coastal upwelling off Oregon. *J. Phys. Oceanogr.*, pp. 164-180.

Ramage, C. S., 1975. Preliminary discussion of the meteorology of the 1972-1973 El Niño. *Bull. Amer. Met. Soc.*, *56*: 234-242.

Rathjen, T. J., 1975. Unconventional Harvest. *Oceanus*, *18*: 21-26.

Reid, J. L., Jr., G. I. Roden, and J. G. Wyllie, 1958. Studies of the California Current System. *Prog. Rep. Calif. Coop. Oceanic. Fish. Invest.*, 1 July, 1956–1 January, 1958, pp. 27-56.

Riley, G. A., 1972. Patterns of production in marine ecosystems. In *Ecosystem Structure and Function*. J. A. Wiens, ed. Corvallis: Oregon State University Press. Pp. 91-112.

Risebrough, R. W., 1971. Chlorinated hydrocarbons. In *Impingement of Man on the Oceans*, D. W. Hood, ed. New York: John Wiley & Sons. Pp. 259-286.

Ryther, J. H., 1969. Photosynthesis and fish production in the sea. *Science*, *166*: 72-76.

Ryther, J. H., 1975. Mariculture, how much protein and for whom? *Oceanus*, *18*: 10-21.

Ryther, J. H., and D. W. Menzel, 1965. *Deep-Sea Res.*, *12*: 199.

Ryther, J. H., E. M. Hulburt, C. J. Lorenzen, and A. Corwin, 1969. *The Production and Utilization of Organic Matter in the Peru Coastal Current*. College Station: Texas A & M Univ. Press.

SCEP, 1970. Study of Critical Environmental Problems: *Man's Impact on the Global Environment*. Cambridge, Mass.: M.I.T. Press, 319 pp.

Shukla, J., 1975. Effect of Arabian sea-surface temperature anomaly on the Indian summer monsoon: A numerical experiment with the GFDL model. *J. Atmos. Sci.*, *32*: 503-511.

Smith, L. J., 1974. Economics of marine pollution. *Oceanus*, *18*: 55-60.

Smith, R. L., 1968. Upwelling. *Oceanogr. Mar. Biol. Ann. Rev.*, *6*: 11-47.

Storer, J. A., and N. Bockstael, 1975. LOS and the fisheries. *Oceanus*, pp. 42-45.

Sverdrup, H. U., 1938. On the process of upwelling. *J. Mar. Res.*, *1*: 155-164.

Thompson, J. D., 1974. The coastal upwelling cycle on a β-plane: hydrodynamics and thermodynamics. Ph.D. thesis, The Florida State University, 141 pp.

Thorade, H., 1909. *Ann. Hydrogr. Bull.*, *37*: 17-34 and 63-76.

Trewartha, G. T., 1961. *The Earth's Problem Climates.* University of Wisconsin Press, 334 pp.

UNESCO, 1970. Contemporary scientific concepts relating to the biosphere. Based on a draft submitted by V. A. Korda and collaborators. Pp. 13-29 in *Use and Conservation of the Biosphere*, National Resources Research, 10, UNESCO, Paris, 272 pp.

Valdivia, J., 1974. Biological aspects of the 1972-1973 "El Niño"— Part 2: The anchovy population. Presented to the IDOE's Workshop on the "El Niño" Phenomenon, Guayaquil, Ecuador, 4-12 December, 1974, 14 pp.

Winstanley, D., 1974. Climatological aspects of drought in the sub-Sahara zone. *Deutsche Geographische Blatter, 51.*

Wooster, W. S., and M. Gilmartin, 1961. The Peru-Chile Undercurrent. *J. Mar. Res., 19*: 97-120.

Wooster, W. S., and J. L. Reid, 1963. Eastern Boundary Currents. In *The Sea, Vol. 2*, M. N. Hill, ed. New York: John Wiley & Sons. Pp. 253-280.

Wooster, W. S., and O. Guillen, 1974. Characteristics of El Niño in 1972. *J. Mar. Res., 32*: 357-404.

Wright, D. J., B. M. Woodworth, and J. J. O'Brien, 1976. A system for monitoring the location of harvestable Coho salmon stocks. *Mar. Fish. Rev., 38*: 1-7.

Wroblewski, J. S., 1976. A model of the spatial structure and productivity of phytoplankton populations during variable upwelling off the coast of Oregon. Ph.D. thesis, The Florida State University, 116 pp.

Wyrtki, K., 1963. The horizontal and vertical field of motion in the Peru Current. *Bull. Scripps Instn. Oceanogr., 8*: 313-346.

Wyrtki, K., 1966. Oceanography of the eastern equatorial Pacific Ocean. *Oceanogr. Mar. Biol. Ann. Rev., 4*: 33-68.

Wyrtki, K., 1973. Teleconnections in the equatorial Pacific Ocean. *Science, 180*: 66-68.

Wyrtki, K., 1974. Equatorial Currents in the Pacific 1950 to 1970 and their relations to the trade winds. *J. Phys. Oceanogr., 4*: 372-380.

Wyrtki, K., 1975. El Niño—the dynamic response of the equatorial Pacific to atmospheric forcing. *J. Phys. Oceanogr., 5*: 572-584.

Wyrtki, K., E. Stroup, W. Patzert, R. Williams, and W. Quinn, 1976. Predicting and observing El Niño. *Science, 191*: 343-346.

7

Climate, Desertification, and Human Activities

William W. Kellogg
Stephen H. Schneider

The climate of the earth has changed many times in the past, and is probably changing even now. One can recognize these changes by observing, for example, the shifting patterns of sea ice and winter snows, changes in the mean position of storm tracks, the growth and decay of mountain glaciers, and the changes in the location and extent of desert regions.

In the past these climate changes were mainly induced by a combination of *natural* influences that researchers are beginning to understand in a general way. However, an increasingly important factor has now been introduced into the complex balance that governs the climate: human activity. One can no longer neglect the often very substantial and growing influence that humans exert by polluting the atmosphere regionally and globally and by altering the character of the earth's surface, in some cases very extensively.

Thus, there are now two important questions concerning climate-desert interaction: what changes may occur in the desert regions when the *global* climate changes because of either natural causes or human activities? What effect does regional modification of desert regions have on the global climate? These two questions are so closely related that it would be misleading to discuss either in isolation. It is recognized, however, that there is the very practical and troublesome problem of trying to identify whether a given change in the geographic scope or in the characteristics of a desert region is the result of a larger general climatic change or the result of the activities of the people living there. This problem will be discussed below.

First, there will be a brief review of the history of climate in the last 20,000 years, and of how the deserts seem to have responded to these changes. The modern record will then be discussed to suggest how short-term fluctuations of the climate may have affected the fringes of deserts. To the people living in these marginal areas, a few consecutive years of drought could mean disaster. Whether or not longer-term changes in global climate are probable seems unimportant to them in comparison with the very real crisis of regional climatic perturbations which may occur in the present, even if these seem minor from a long-term global perspective.

There will then be a review of some current theories of how human activities may be modifying the climate of the world in general and of desert regions in particular. Although the deserts (defined here as places where annual precipitation is less than 200 mm per year) do not occupy more than about one-quarter of the global land area (Köppen, 1936), they probably play an important role in determining the earth's heat and water balances. If human activities, therefore, are altering the size of deserts and their margins, such changes could influence more than just the immediate region involved.

Many of these physical interactions will have to be described in very qualitative terms, because enough is not yet known about our planet and its climate to say definitively how a given alteration might affect the state of the atmosphere, oceans, land, and ice—the basic components of the climate system. In addition, while "models" of the system are constantly being built and improved, they must still be considered too primitive to yield precise answers concerning the physical interactions of the oceans, the atmosphere, and the land. Nevertheless, their results often raise issues that must be addressed, and several of these are considered subsequently.

Deserts of the Past

Deserts tend to be situated in the subtropics[1] (the region between the humid tropics and temperate mid-latitude zones) of the major continents, including Australia, as well as in parts of the polar regions including most of Antarctica and the north coast of Greenland. One might argue that some deserts exist over large

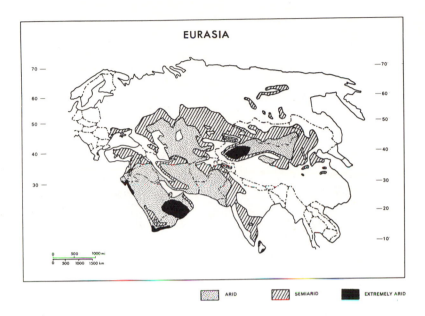

Figure 7.1a. The distribution of extremely arid, arid, and semiarid lands: Eurasia. Reprinted with permission of AAAS and of H. E. Dregne as editor of *Arid Lands in Transition* (Washington, D.C.: AAAS, 1970).

parts of the oceans in the sense that little rainfall occurs in certain of these regions. The deserts occur primarily because the large-scale circulation subsystems of the atmosphere are either unable to transport much moisture to them, or they are unable to sustain the processes that can release the atmospheric moisture in the form of precipitation. Atmospheric motion, or circulation, then, is a crucial factor in the existence of these major deserts.

Figures 7.1a and 7.1b show the location of the major deserts of the world. The largest ones are in North Africa, Australia, North America, and Asia (both in northwest India and in the high plateau region). It is significant that most of these deserts appear to have once been either prairie land or savannah, or they were considerably less extensive area-wise only 4,000 to 6,000 years ago (Lamb et al., 1966; Kellogg, 1976). What was different then? One major difference seems to have been that the earth was generally warmer. Thus, one must try to find some linkage between the earth's average temperature and atmospheric circulation systems.

With respect to the period of the last major glaciation

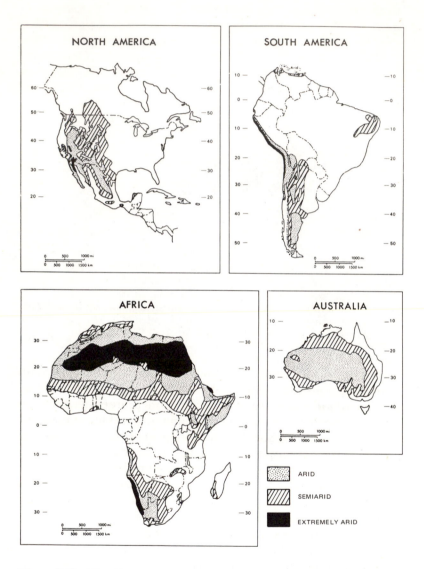

Figure 7.1b. The distribution of extremely arid, arid, and semiarid lands: North and South America, Africa, and Australia. Reprinted with permission of AAAS and of H. E. Dregne as editor of *Arid Lands in Transition* (Washington, D.C.: AAAS, 1970).

(the "Wisconsin Period"), which reached its climax about 18,000 to 20,000 years ago, the record becomes more difficult to decipher. However, it is generally believed that in that period precipitation was less over the continents generally (Williams et al., 1974; Lamb, 1974; Hammond, 1974; Newell et al., 1975), and that the subtropical regions of the major continents were more arid than now. Monsoonal rain belts are also believed to have been considerably reduced from today's levels.

While a *full* discussion of how the large-scale circulation patterns shifted to account for these changes in mean rainfall patterns is beyond the scope of this chapter—and, quite frankly, the state of the art—a partial and somewhat simplified explanation can be made in terms of the response of the basic atmospheric circulation systems.

The tropical regions are heated by the sun vastly in excess of the heating of the colder polar areas, and the atmosphere and oceans tend to make up this heating imbalance by transporting heat from the excessively heated tropics to the poles. During this process, the wind and ocean currents that determine the climate are generated. These circulation patterns are shown schematically in figure 7.2. Moreover, there is a direct meridional circulation cell, the Hadley cell, that has a rising arm in the tropics and a descending arm in the subtropics, as can be seen in figure 7.2. In the rising part, water vapor is transported vertically, thereby expanding, a process by which it is cooled, condensed, and finally falls as precipitation. In the descending arm, however, the opposite effect is produced: the drier air sinks, is compressed, and tends to produce a region of dry air that inhibits the formation of precipitation, despite the possible existence of water below. Recall that even some ocean regions are deserts with respect to precipitation amounts.

The vigor and extent of the Hadley circulation depends on the equator-to-pole temperature difference, a difference which drives the circulation. Indeed, the atmosphere behaves as a heat engine that is "fueled" by the difference between temperatures in the equatorial zone and the polar regions. However, it is not only the solar radiation heating gradient that causes the temperature contrasts between equator and poles. For example, the earth and its atmosphere radiate energy proportional to the fourth power of its temperature. Because the earth's radiative temperature is roughly 255°K (−18°C), it radiates energy to space in the "infrared" portion of the electromagnetic spectrum. Although both tropical and polar regions of the earth lose infrared radiation energy to space, the

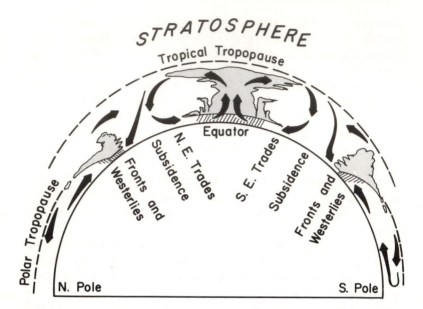

Figure 7.2. Schematic of the mean meridional circulation in the troposphere.

amount of infrared radiation escaping in polar regions exceeds the amount of solar energy absorbed, or a "negative radiation balance," whereas the amount of solar energy absorbed in tropical regions exceeds the amount of infrared radiation escaping to space, or a "positive radiation balance." On a *global* basis, the amount of solar energy absorbed is balanced by the amount of infrared energy lost, and this equilibrium maintains the earth's mean temperature at a nearly constant value (see, for example, Schneider and Kellogg, 1973).

However seemingly small, changes in that balance could have important consequences for desert regions in particular as well as for the earth's climate in general. In fact, even small changes due to slow changes in the character of the earth's orbit around the sun coupled with the direction of the earth's axis of rotation, on a time scale of 10,000 years or more, may account for major climate change like ice ages (Kukla, 1975).

During a period when the earth is generally a little warmer, it seems likely that the polar regions warm up more than the tropics (SMIC, 1971; Sellers, 1973; Wetherald and Manabe, 1975). In such a case there is less temperature difference between the equator and the poles, the mid-latitude westerlies (figure 7.2) are generally displaced poleward, and the Hadley circulation, we presume, is less concentrated and occupies a larger band on each side of the equator. This, in fact, occurs each year in the summer hemisphere. On the other hand when it is generally colder the Hadley cell is restricted to a narrower latitude band, and the downward arm more vigorously suppresses precipitation in the parts of the subtropics under its domination.

Most of Asia and East Africa is influenced by the Asian monsoon, a related circulation system. In summer this monsoon circulation draws moist air from the Indian Ocean up the slopes of the Tibetan Plateau, bringing rain to most of southern Asia. For reasons not entirely understood, but probably related to circulation pattern adjustments, the summer monsoons seem to have been stronger during the warm periods and weaker during the cold glacial periods (Williams et al., 1974; Bryson, 1974). This must have had a major influence on the aridity of such climatically marginal regions as northwest India.

Archaeological records, the history of lake levels, the past distribution of species of trees and shrubs as ascertained by their patterns of deposits in lake beds and peat bogs, the patterns of ancient sand dunes, and other similar information, allow the researcher to piece together changing patterns of past deserts, prairies, and forests. We see, for example, that 4,000 to 6,000 years ago, when conditions were generally warmer than now, parts of North Africa now encompassed by the Sahara Desert were mostly prairie grassland. Here some inhabitants hunted game and some lived on snails, most of which have long since disappeared from the region (Butzer, 1966; Street and Grove, 1976; Lubell et al., 1976). The Fertile Crescent (including the valley of the Tigris and Euphrates Rivers) was indeed "fertile" then, as was the region of the eastern Mediterranean (Lamb, 1972). The Rajasthan and Thar Deserts of northwest India supported cattle and goats, and were the sites of cities long since abandoned to desert sands (Bryson and Baerreis, 1967). For the deserts of southwestern North America the record is not as clear, but some areas, such as New Mexico, may have experienced more rain than now, especially during the later part of the Altithermal (Martin and Mehringer, 1965; Malde, 1974).

What Makes It Rain in a Desert?

Having discussed the influence of large-scale global circulation patterns on desert regions, another set of factors connected with the vertical stability of the atmosphere must be considered. Vertical stability is determined by the vertical distribution of temperature and moisture in the air mass over a region. Conditions for the development of cumulus clouds and thunderstorms are most favorable when the lower layers contain considerable moisture, and heating during the day causes the air to become sufficiently buoyant to rise in "convective plumes." The rising air is then cooled by expansion, and at a level called "the lifting condensation level" moisture begins to condense into cloud droplets. As the air continues to rise, its liquid water amount increases. The effect of this condensation is to release the "latent heat" of the water vapor—which was used originally to cause evaporation—and to add further to the buoyancy of the convective plume. Although such a rising parcel becomes colder than when it began its ascent, it maintains its buoyancy as long as it remains warmer and, therefore, less dense, than the surrounding air at the same altitude. When conditions are favorable, that is, when the atmosphere is vertically unstable, this process can continue until the plume grows into a fully developed thunderstorm with its associated precipitation.

There is another process taking place in the upper part of such a cloud, a process that enhances its chances for precipitation. Since the upper part of such a cloud can be influenced artificially, a brief explanation will be made. After a cloud droplet in the plume (or updraft) reaches the freezing level, it is cooled below the freezing point of water, that is, it becomes "supercooled." However, this does not mean that it will immediately freeze, since very pure water can be supercooled to about $-40°C$ before freezing. In order to freeze it must either encounter or it must already contain a special kind of particle called a "freezing nucleus" which can start the crystallization of ice. The measure of the "efficiency" of different kinds of freezing nuclei is the temperature at which they will initiate freezing in a supercooled droplet. In some air masses there is a lack of efficient freezing nuclei; in others an abundance. A recognized method of deliberately influencing convective storms is to add more of such nuclei (usually silver iodide particles). How do such freezing nuclei have an effect on precipitation?

The following theory is the one that is usually advanced, even though it has been apparently difficult to prove by actual field experiments. It is known that when an ice particle is created in a cloud of supercooled droplets it grows rapidly at the expense of the droplets, the explanation being that the "vapor pressure" at its surface is appreciably lower than the vapor pressure around the liquid droplets, so water molecules diffuse preferentially to the ice particle. The larger ice particle then starts to fall, and in doing so collides with supercooled water droplets and thereby grows even faster. Thus, by the creation of a relatively few ice crystals a cloud of supercooled droplets can produce precipitating particles much more quickly than the droplet-droplet interactions alone could do. This has led to the common belief that *some* freezing nuclei must be required before a large cumulus cloud can effectively rain, even after it has extended in altitude above the freezing level.

In summary, the following conditions are most favorable for rain: The air must be sufficiently unstable in the lower layers to allow convective plumes to form. There must be enough moisture in the air to create water droplets in the updraft. For the efficient production of rain, many clouds must build up to the freezing level *and* contain some freezing nuclei of sufficient efficiency. Furthermore, most of the rain in tropical areas, as in mid-latitudes, is associated with "synoptic scale" disturbances of roughly 1,000 km length and a few days' duration. These disturbances are embedded in the larger-scale circulation patterns such as those of the summer monsoon.

The first of these conditions—the initial creation of the plume (or updraft)—is often likely to be attained as air passes upward over a mountain where either the warm slopes can effectively heat the air or where the air is physically forced upward. This condition is least likely to occur where there is a general downward motion of the air aloft, i.e., subsidence. Such subsidence of dry air causes a warmer, more stable "inversion" layer, a situation in which temperature increases with altitude. Such an inversion layer hinders the rise of a convective plume (an inversion layer aloft is sometimes referred to in this context as a "lid" on the convective processes).

The rate of heating or cooling caused by layers of particles ("aerosols") that may absorb sunlight and infrared radiation is another process that may influence the stability of the lower layers. Visualize the layer of dust or haze that commonly occurs over dry regions, a layer that typically extends from the earth's surface up to some rather distinct transition altitude from dusty to clear air. During the daytime, sunlight will be both absorbed and scattered by such a layer, warming the particles and the air in the upper part, and preventing some solar radiation from reaching the surface. Such daytime warming aloft with a relative cooling of the surface will increase the stability and thereby decrease convective activity. At night the effect of infrared radiation to space from the upper part of the dust or haze layer may cool it slightly, but this effect is probably less significant than the daytime heating. Furthermore, rains are rarely initiated at night in a desert region.

Human Influence on Deserts

Deserts are situated where the large-scale global circulation patterns either deny moisture to a region or where subsiding air associated with those patterns puts a lid on rain-producing convection. In addition, there are many localized processes at work as well, such as the warming—or lack of warming—of the air in contact with the ground, and such as the degree of the availability of freezing nuclei. The question now is whether human activities can influence these desert-related processes. The answer is a qualified "yes."

Consider first the regional or local processes. Much has been written about the implications of livestock overgrazing in marginally vegetated desert regions. There seems to be little doubt that such practices can indeed denude the vegetation from large areas such as the Sahel (Bryson, 1973, 1974; Glantz, 1976), the Negev-Sinai (Otterman, 1974), and the Rajasthan (Bryson and Baerreis, 1967). Satellite photographs (Rapp, 1975; Wage, 1974; Otterman, 1974) have shown that fences or political boundaries mark definite lines separating the relatively light areas of exposed ground from areas where control of grazing had left the darker areas of standing vegetation or debris more or less intact. It becomes clear from such photographs that this form of human activity has increased the surface reflectivity (albedo) of the ground. Otterman (1977) has suggested that the observed increase in surface albedo due to heavy grazing pressures is not only a consequence of the removal of the vegetation cover, but also is due to the reduction of the leaf litter, also referred to as debris, between the plants. That litter, which appears darker than most soils or sands, helps to form a darkened crust which can be, and often is, broken up by the trampling of

(1972) and Bolin and Charlson (1976). It is suggested that the warming would be due primarily to the addition of carbon dioxide to the atmosphere as we burn the carbon, which is currently stored in the ground as coal, petroleum, natural gas, and peat. The concentration of carbon dioxide has been increasing at a rate of almost 1% per year since about 1958, when good observations in both hemispheres began to be made continuously, or nearly so, from a number of stations. It is anticipated that the rate of the burning of fossil fuels will continue to increase at about 4% per year until at least 2000 A.D. It is also anticipated that about half of all the carbon dioxide produced by then will have remained in the atmosphere with the other half having been taken up by the oceans and forests around the world. It is important to note that with respect to the forests they should in fact grow at a faster rate as the carbon dioxide content increases (SMIC, 1971; Machta and Telegadas, 1974; Machta, 1973; Bacastow and Keeling, 1973; Mitchell, 1972, 1974; Broecker, 1975). Therefore, the wholesale cutting of tropical forests resulting in still more carbon dioxide remaining in the atmosphere may become a crucial factor with respect to global temperatures (Garrels et al., 1976).

The connection between carbon dioxide, currently at an atmospheric concentration of almost 330 parts per million by volume (ppmv), and the global mean surface temperature arises because carbon dioxide, while relatively transparent to solar radiation, absorbs quite strongly in that part of the infrared spectrum between 8 and 17μ m known as "the infrared window," where water vapor, the other major infrared absorber, has very little absorption. Thus, some infrared radiation from the surface that would otherwise have escaped to space and thereby have cooled the surface is intercepted by the carbon dioxide. This energy is then reradiated both upward *and downward*. The infrared radiation in the carbon dioxide absorption band that is finally lost to space from the upper levels of the atmosphere is emitted by a much colder region. Since the amount radiated is a very strong function of temperature, the net result of this process is a warmer troposphere and a colder stratosphere, which is the region above about 12 km (for example, SMIC, 1971; Schneider and Kellogg, 1973).

The amount of warming at the surface has been estimated by a number of independent models of the atmosphere that take account of the change in infrared absorption due to a change in carbon dioxide, together with a number of other factors such as the tendency of the atmosphere to have a constrained vertical temperature profile, or lapse rate, and a constant relative humidity. The most detailed calculation to date is Manabe and Wetherald (1975). These various model calculations have been reviewed by Schneider (1975) and Budyko (1976), and the consensus of state-of-the-art theories is that for a doubling of the carbon dioxide content to about 650 ppmv (probably to be attained by the middle of the next century) the mean *global* surface temperature rise will be on the order of 2-3°C (4-6°F). It is quite possible, however, that processes not accounted for by these models could raise or lower this estimate by severalfold. On a shorter time scale, by 2000 A.D., when the carbon dioxide content of the atmosphere will be approaching 400 ppmv, the global surface temperature rise is estimated to be about 1°C. Note that these are all estimates for the mean global temperature change. At middle and high latitudes such changes will probably be greater.

There are other human activities that could affect the global or hemispheric heat balance of the atmosphere, the most notable being the following: the injection of particles or of gases such as sulfur dioxide that turn into particles; the injection of other trace gases such as the chlorofluoromethanes used in aerosol spray cans and refrigerators, gases that absorb in the infrared window (Ramanathan, 1975; NAS, 1976); the direct addition of heat locally referred to as "thermal pollution." The latter, however, is not likely to have a large *global* impact in the next 100 years or so. All of these additions to the atmosphere are very likely to *warm* the lower atmosphere (Kellogg, 1975, 1977; Kellogg and Schneider, 1974). Even the particles could serve to warm the climate (Kellogg et al., 1975), although this is a controversial point, since some people have argued that the particles could have recently cooled the Northern Hemisphere (Rasool and Schneider, 1971; Yamamoto and Tanaka, 1972; Bryson, 1972; Bolin and Charlson, 1976).

To place this estimate of an artificially induced global warming of about 1°C by 2000 A.D. into proper perspective with respect to natural climate changes that have occurred during the past millennium, and recalling that the changes at middle and high latitudes could well be on the order of four times larger, it is relatively safe to say that the kind of warming discussed here would lead to a climate warmer than at any time since before the major cooling trend that occurred 3,000 to 5,000 years ago. The mean temperatures during the Little Ice Age, that lasted from roughly 1600 to 1850 A.D., may have been as much as a few °C colder than now (NAS, 1975). This discussion relates to the consequences of adding another 1°C to the present mean temperature, which already appears to be as warm or warmer than at any time in the past 1,000 years. One must be cautious, however, in view of the fact that it is difficult to say much about mean global temperatures before the middle of the nineteenth century, when quantitative worldwide meteorological observations were just beginning (Schneider and Mass, 1975). In fact, it is difficult to be very precise even now (van Loon and Williams, 1976a, 1976b).

If the current upward trend in human population and in energy production continues beyond 2000 A.D., the mean global temperature should continue to rise. As human impact on the atmosphere grows, it will become increasingly unlikely that any natural influence will be strong enough *on such a short time scale* to counteract it. Between the beginning and the middle of the twenty-first century, the earth could become as warm as the period between about 8,000 and 4,000 years ago, known as the Altithermal. There now exists almost universal agreement among paleoclimatologists that the Altithermal was indeed generally warmer worldwide, and that the change was greater at high latitudes than in the tropics (for example, see Lamb, 1972). Since it is possible that we could now in essence be "rolling back the clock" to a warmer earth, it becomes important to know what the desert regions of the world were like during the Altithermal.

There is no assurance, however, that a warmer twenty-first century earth would be exactly like the Altithermal, since the way in which the extra heat would be supplied to the lower atmosphere is not necessarily the same as that which existed during the Altithermal. Whereas the main reason for the future warming is expected to be the entrapment of infrared radiation by added carbon dioxide, a possible explanation for the Altithermal is that the earth's elliptical orbit around the sun and the direction of the earth axis 4,000 to 6,000 years ago were such that the summer and fall seasons enjoyed more solar radiation in the Northern Hemisphere. Thus, the extra land area of the Northern Hemisphere could have allowed the earth to respond to the extra summertime heating by rolling back the polar ice and delaying the onset of the first snowstorms (Kukla, 1975). Because of these differences in the causes of heating, one cannot assume that the all-important circulation patterns will again be like those of that former warm period.

Nevertheless, the Altithermal is the only evidence that exists, other than theoretical models, concerning the characteristics of a few °C warmer climate, and it is thus instructive to show how different it was from today's climate. The picture that emerges as one pieces together evidence gathered from various parts of the world by paleoclimatologists is as follows: the earth, and especially the subtropics, was, in general, wetter during the Altithermal period. Previously it was described how many of the present desert regions were habitable until about 4,000 years ago, while some continental regions at mid-latitudes, such as central North America and Scandinavia, were somewhat drier.

This picture of a planet with more rainfall in the subtropics and in some mid-latitude continental areas is apparently supported in general by an experiment, mentioned earlier, with a Geophysical Fluid Dynamics Laboratory general circulation model. This model, designed by Manabe and Wetherald (1975), included a simple noncirculating ocean as well as a quasi three-dimensional atmosphere. When this model was run with either 2% more solar radiation (Wetherald and Manabe, 1975) or a doubling of carbon dioxide (Manabe and Wetherald, 1975), the surface temperature increased. In both experiments the entire hydrologic cycle was more vigorous in that more water was evaporated at low latitudes, transported by the atmosphere to middle latitudes, and deposited there as rain or snow. Another result of the experiments was that, while there was only a small increase of precipitation in the tropical belt as a whole, there was a kind of major "land-sea breeze effect," and the model continent apparently warmed the moist air passing over it enough to cause relatively more rainfall over the land than over the ocean in the tropics (Manabe, personal communication). However, this model did not include the real continents and mountain ranges of the world and so detailed conclusions about just where the increased precipitation would occur cannot be drawn.

Concluding Comments

One must draw the conclusion from the preceding discussion that human activities have the ability to modify the desert areas around the world, though to date such modifications have for the most part been inadvertent. For example, on a local or regional scale domesticated livestock can modify the albedo of the surface of desert margins by trampling and overgrazing, and can thereby also change the availability of freezing nuclei. Both of these factors have been suggested as capable of decreasing precipitation where it is needed most. However, on a global scale, with all other factors constant, human activities may be moving the climate toward a regime that will bring more rainfall to the subtropical deserts of the world.

The arguments presented in this chapter have been mainly theoretical and have been based on a somewhat incomplete understanding of all the physical processes and interactions that govern the climate of deserts. Also a number of the arguments have been based on models of the atmosphere and of its radiation balance, both of which must still be considered oversimplified relative to reality. However, with some concrete evidence to back the arguments presented, evidence which shows conclusively that large parts of some semidesert areas have been modified by human activities, the theoretical extension of the argument is not unreasonable. Such arguments, however, should not be extrapolated out of proportion to the existing evidence.

A difficulty in dealing with any subject relating to the interactions between human activities and the climate is the very dominant role played by nature, a role that cannot yet be well predicted. To determine, for example, the human influence on the rainfall and/or climate of a region—or of the globe—through observations, such influences would have to be larger than the natural fluctuations so that one would be able to distinguish the climatic "signal" of human influence from the background "noise." Thus, while the total impact of mankind on the climate will probably become significant and identifiable by 2000 A.D., it is hard to demonstrate whether humans have *already* had an impact— though it is the belief of the authors of this chapter that there may very well have been an impact.

The view of the future, however, does not appear to be utterly bleak. While the practice of overgrazing marginal lands must be considered as detrimental on ecological grounds alone, there is yet no definitive measure that can determine just how much grazing an ecological system can bear, especially when one takes account of the natural variability of the rainfall that supports the vegetation. The greatest risk of course is to allow herds to expand in good rainfall years beyond the carrying capacity of the *poorest* years, for then overgrazing pressures can lead to an amplified catastrophe, as was the case in the Sahel between 1968 and 1973. However, if the global rainfall patterns shift in a way that is favorable to these low rainfall marginal areas, they will presumably be able to withstand more pressure, as was generally the case in the Sahel between 1952 and 1967.

Yet for argument's sake, if an increase in rainfall in subtropical regions were to occur and if this were accompanied by a simultaneous deterioration of agriculture in the mid-latitude grainbelts, such as Budyko (1976) suggests would occur with a planetary warming, then the resulting situation would not necessarily be a boon to the subtropical countries. If the large surpluses produced by mid-latitude granaries were to be sharply reduced, world food prices in all probability would increase sharply as the availability of food for export would be reduced. The impact on food-deficit nations, given such a scenario, could be disastrous. Finally, arguing the other extreme (that the climate could conceivably cool from either natural or human effects), the situation that we consider more probable could be reversed. The implications for the subtropical countries of either scenario is fraught with problems and significant uncertainties. The problem often is sudden change, regardless of the direction of change.

Thus, because of the existing level of interdependence among nations due to the international food trade and due to the awareness that world food production fluctuates along with the climate, maintenance of an adequate world food reserve and distribution system—a "Genesis strategy" (Schneider and Mesirow, 1976)—seems essential as a hedge against the growing prospect of climate change, regardless of whether the global atmosphere is warming or cooling.

Our suspected most-likely future scenario presented in this chapter is not to be viewed as a firm forecast. The scenario might hold out hope for those who live on the fringes of desert regions in that they will not necessarily face unfavorable climatic trends in the long term. Still, such a possibility should not be allowed to weaken a commitment to protect the fragile desert environments from probable irreversible damage by maintaining herds at a level below carrying capacity in the *worst* years, and by maintaining adequate reserve capacity to survive the frequent droughts that will occur in such climatically marginal areas because such events are part of such regions' climate regimes.

Acknowledgments

The authors appreciate the constructive criticism of this chapter by their colleagues, Drs. Edward J. Zipser and Michael H. Glantz.

Note

1. Some deserts occur where a mountain range removes some of the water vapor from the air as it is forced by the prevailing wind up and over the range, leaving a dryer "rain shadow" in its lee. Such deserts are, relatively, not extensive and, barring major changes to the large-scale circulation patterns (sufficient to change the prevailing wind direction), they are likely to persist as long as does the mountain range that creates them. These deserts will not be discussed.

References

Bacastow, R., and C. D. Keeling, 1973. Atmospheric carbon dioxide and radiocarbon in the natural carbon cycle: II. Changes from A.D. 1700 to 2070 as deduced from a geochemical model. Chapter in *Carbon and the Biosphere*, G. M. Woodwell and E. V. Pecan, eds. (Brookhaven Lab.), U.S. Atomic Energy Comm. CONF-720510 (available from Natl. Tech. Info. Service).

Bolin, B., and R. J. Charlson, 1976. On the role of the tropospheric sulfur cycle in the shortwave radiative climate of the earth. *Ambio*, 5: 47-54.

Broecker, W. S., 1975. Climatic change: Are we on the brink of a pronounced global warming? *Science*, 189: 460-463.

Bryson, R. A., 1972. Climate modification by air pollution. In *The Environmental Future*, N. Polunin, ed. New York: Macmillan. Pp. 133-154.

Bryson, R. A., 1973. Climatic modification by air pollution. II: The Sahelian effect. Report no. 9, Madison Institute for Environmental Studies, University of Wisconsin.

Bryson, R. A., 1974. A perspective on climatic change. *Science*, 184: 753-760.

Bryson, R. A., and D. A. Baerreis, 1967. Possibilities of major climatic modification and their implications: Northwest

India, a case for study. *Bull. Amer. Meteor. Soc.*, 48: 136-142.

Budyko, M. I., and K. Ya. Vinnikov, 1976. Global warming. *Meteorologiya I Gidrologiya* (Moscow), 7: 16-26. Translated by NOAA Language Services Division.

Budyko, M. I., 1976. On present-day climatic changes. Submitted to *Tellus*.

Butzer, K. W., 1966. Climatic changes in the arid zones of Africa during early to mid-Holocene times. In *World Climate from 8000 to 0 B.C.*, J. S. Sawyer, ed. London: Royal Meteorological Society. Pp. 72-83.

Charney, J. G., 1975. Dynamics of deserts and drought in the Sahel. *Q. J. Roy. Meteor. Soc.*, 101: 193-202.

Charney, J., P. H. Stone, and W. J. Quirk, 1975. Drought in the Sahara: A biophysical feedback mechanism. *Science*, 187: 434-435.

Eckholm, E. P., 1975. Desertification: A world problem. *Ambio*, 4: 137-145.

Garrels, R. M., A. Lerman, and F. T. Mackenzie, 1976. Controls of atmospheric O_2 and CO_2: Past, present, and future. *American Scientist*, 64: 306-316.

Glantz, M. H., ed., 1976. *The Politics of Natural Disaster: The Case of the Sahel Drought.* New York: Praeger Publishers, 336 pp.

Hammond, A. L., 1976. Paleoclimate: Ice Age Earth was cool and dry. *Science*, 191: 455.

Kellogg, W. W., 1975. Climate change and the influence of man's activities on the global environment. Chapter in *The Changing Global Environment*, S. F. Singer and D. Reidel, eds. Dordrecht-Holland. Pp. 13-23.

Kellogg, W. W., 1977. Global influences of mankind on the climate. Chapter in *Climate Change*. J. Gribbin, ed. New York: Cambridge University Press.

Kellogg, W. W., J. A. Coakley, and G. W. Grams, 1975. Effect of anthropogenic aerosols on the global climate. *Proc. WMO/ IAMAP Symp. on Long-Term Climatic Fluctuations*, Norwich, U.K. WMO Document 421, Geneva, pp. 323-330.

Kellogg, W. W., and S. H. Schneider, 1974. Climate stabilization: For better or for worse? *Science*, 186: 1163-1172.

Köppen, W., 1936. Der geographische system der klimate. In *Handbuch der Klimatologie*, Köppen and Geiger, Band I, Teil C, Borntraeger, Berlin.

Kukla, G. J., 1975. Missing link between Milankovitch and climate. *Nature*, 253: 600-603.

Lamb, H. H., 1972. *Climate: Present, Past and Future.* Vol. I: Fundamentals and Climate Now. London: Methuen and Co., 613 pp.

Lamb, H. H., 1974. Climates and circulation regimes developed over the northern hemisphere during and since the last ice age. *Proc. IAMAP/WMO Symp. on Physical and Dynamic Climatology,* Leningrad, August 1971, World Meteor. Org. No. 347, Gidrometeoizdat, Leningrad, U.S.S.R., pp. 233-261.

Lamb, H. H., R. P. W. Lewis, and A. Woodroffe, 1966. Atmospheric circulation and the main climatic variables between 8000 and 0 B.C.: Meteorological evidence. *Proc. Intl. Symp. on World Climate 8000 to 0 B.C.,* Roy. Meteor. Soc., London, pp. 174-217.

Lubell, D., F. A. Hassan, A. Gautier, and J.-L. Ballais, 1976. The Capsian Escargotieres. *Science,* 191: 910-919.

Machta, L., 1973. Prediction of CO_2 in the atmosphere. Chapter in *Carbon and the Biosphere,* G. M. Woodwell and E. V. Pecan, eds. Brookhaven Lab., U.S. Atomic Energy Comm. CONF-720510 (available from Natl. Tech. Info. Service), pp. 21-31.

Machta, L., and K. Telegadas, 1974. Inadvertent large-scale weather modification. Chapter 19 in *Weather and Climate Modification,* W. N. Hess, ed. New York: John Wiley and Sons. Pp. 687-726.

Malde, H. E., 1974. Environment and man in arid America. *Science,* 145: 123-129.

Manabe, S., and R. T. Wetherald, 1975. The effects of doubling the CO_2 concentrations on the climate of a general circulation model. *J. Atmos. Sci.,* 32: 3-15.

Martin, P. S., and P. J. Mehringer, Jr., 1965. Pleistocene pollen analysis and biogeography of the southwest. In *The Quaternary of the United States,* H. E. Wright and D. G. Frey, eds. Princeton, N.J.: Princeton University Press. Pp. 433-451.

Mitchell, J. M., Jr., 1972. The natural breakdown of the present interglacial and its possible intervention by human activities. *Quaternary Res.,* 2: 436-445.

Mitchell, J. M., Jr., 1974. The global cooling effect of increasing atmospheric aerosols: Fact or fiction? *Proc. IAMAP/WMO Symp. on Physical and Dynamic Climatology,* Leningrad, August 1971, World Meteor. Org. No. 347, Gidrometeoizdat, Leningrad, U.S.S.R., pp. 304-319.

NAS, 1975. *Understanding Climate Change; A Program for Action.* U.S. Comm. for GARP, National Academy of Sciences,

Washington, D.C.

NAS, 1976. *Halocarbons: Environmental Effects of Chlorofluoromethane Release*. Comm. on Impacts of Stratospheric Change, National Academy of Sciences, Washington, D.C.

Newell, R. E., G. F. Herman, S. Gould-Stewart, and M. Tanaka, 1975. Decreased global rainfall during the past ice age. *Nature*, 253: 33-34.

Otterman, J., 1974. Baring high-albedo soils by overgrazing: A hypothesized desertification mechanism. *Science*, 186: 531-533.

Otterman, J., 1977. Anthropogenic impact on the albedo of the earth. *Climatic Change*, 1:2.

Ramanathan, V., 1975. Greenhouse effect due to chlorofluorocarbons: Climatic implications. *Science*, 190: 50-52.

Rapp, A., 1975. Soil erosion and sedimentation in Tanzania and Lesotho. *Ambio* 4 (4), pp. 15-103.

Rasool, S. I., and S. H. Schneider, 1971. Atmospheric carbon dioxide and aerosols: Effects of large increases on global climate. *Science*, 173: 138-141.

Schneider, S. H., 1975. On the carbon dioxide-climate confusion. *J. Atmos. Sci.*, 32: 2060-2066.

Schneider, S. H. (with L. E. Mesirow), 1976. *The Genesis Strategy: Climate and Global Survival*. New York: Plenum Publishing Corp., 419 pp.

Schneider, S. H., and C. Mass, 1975. Volcanic dust, sunspots, and temperature trends. *Science*, 190: 741-746.

Schneider, S. H., and W. W. Kellogg, 1973. The chemical basis for climate change. Chapter 5 in *Chemistry of the Lower Atmosphere*, S. I. Rasool, ed. New York: Plenum Publishing Corp. Pp. 203-250.

Schnell, R. C., 1976. Biogenic ice nuclei: Part I. Terrestrial and marine sources. *J. Atmos. Sci.*, 33: 1554-1564.

Sellers, W. D., 1973. A new global climatic model. *J. Appl. Meteor.*, 12: 241-254.

SMIC Report, 1971. *Inadvertent Climate Modification: Report of the Study of Man's Impact on Climate*. Cambridge, Mass.: M.I.T. Press.

Street, F. A., and A. T. Grove, 1976. Environmental and climatic implications of late Quaternary lake-level fluctuations in Africa. *Nature*, 261: 385-390.

Vali, G., M. Christensen, R. W. Fresh, E. L. Galyan, L. R. Maki, and R. C. Schnell, 1976. Biogenic ice nuclei: Part II. Bacterial sources. *J. Atmos. Sci.*, 33: 1565-1570.

van Loon, H., and J. Williams, 1976a. The connection between trends of mean temperature and circulation at the surface: Part I. Winter. *Mon. Wea. Rev.*, 104: 365-380.

van Loon, H., and J. Williams, 1976b. The connection between trends of mean temperature and circulation at the surface. Part II. Summer. *Mon. Wea. Rev.*, 104: 1003-1011.

Wade, N., 1974. The Sahelian drought: no victory for western aid. *Science*, 185: 234-237.

Wetherald, R. T., and S. Manabe, 1975. The effects of changing the solar constant on the climate of a general circulation model. *J. Atmos. Sci.*, 32: 2044-2059.

Williams, Jill, R. G. Barry, and W. M. Washington, 1974. Simulation of the atmospheric circulation using the NCAR global circulation model with ice age boundary conditions. *J. Appl. Meteor.*, 13: 305-317.

Yamamoto, G., and M. Tanaka, 1972. Increase of global albedo due to air pollution. *J. Atmos. Sci.*, 29: 1405-1412.

8

Desertification and Population: Sub-Saharan Africa

Helen Ware

The more fragile the environment, the more delicate the balance between population and ecological resources. Hence, the linkage between desertification and population is a justified concern for research lest the growth of human populations may prove to have been a major contributory factor in increasing desertification in sub-Saharan Africa. Unfortunately, many commentators have gone much further than this to argue that unchecked population growth in the region will inevitably lead to irreversible disaster. *World Population and Food Supplies*, which, although personally attributed to Lester Brown, was one of the four official United Nations background papers presented to the World Population Conference in Bucharest, is an excellent example. This pronounces that due to

> denudation and deforestation, the Sahara desert has begun to move southward at an accelerated rate all along the 3,500 mile southern fringe, stretching from Senegal to northern Ethiopia. An "in-house" study undertaken by the United States Agency for International Development in August 1972 indicated that the desert is moving southward at up to 30 miles per year, depending on where it is measured. There is now a pressing need to address and alleviate the causes of ecological stress in the region

> which facilitates the southward movement of the Sahara.
> Failure to do so means the Sahara may engulf much of
> central Africa in a matter of years destroying a significant
> slice of the continent's food producing capacity. . . .
> Above all it calls for the launching of efforts to slow and
> stabilize population growth in the region (Brown, 1974).

This vision of man's capacity to advance the Sahara to the banks of the River Congo within a century would be humorous were its false diagnosis not so pernicious.

If there has been a major alteration in climatic conditions due to a change in global air pressure belts, then population growth may possibly have aggravated the problem, but certainly did not cause it (Lamb, 1973; Poursin, 1974). Indeed, the most reasonable conclusion appears to be that although "years of high or low rainfall tend to be bunched together in small clusters of two or three years and in longer periods of ten or fifteen years . . . there is no indication of any long continued upward or downward trend in rainfall nor is there any obvious cyclic periodicity" (Grove, 1973). The 1970-73 drought was not the worst ever in the Sahel. There are records of five-year droughts affecting the Niger Bend as far back as the eighteenth century (Schove, 1973). Most of the savannah people believe that the drought of 1913 was the worst within living memory, and certainly the loss of human—though possibly not animal—life was much worse then than in the latter, much more publicized drought. All the evidence, where there are rainfall records, suggests that the drought of the 1910-15 period was worse than that of the 1970s in the savannah farming zone though probably not in the northern Sahel and Mauritania (Roche, 1973).

The true problem is that, whereas formerly the proportion of the human population which died during major droughts was sufficient to impede all but the slowest of long-term population growth rates, this is no longer the case. In the early 1970s the efforts of the peoples of the six Sahelian countries themselves together with international aid served to ensure that at an absolute, and most improbable, upper limit 100,000 people who would not otherwise have died, succumbed to the effects of famine (Center for Disease Control, 1973). All such deaths represent a tragic waste, but even this maximum represents only a rise of one point per thousand in the overall death rates of the region during the four years of the drought (Caldwell, 1975a). Whereas in 1913 the very existence of the Sahelian drought passed virtually unperceived in the outside world, 60 years later in 1973 well over .5 million tn of grain were brought into the region and distributed. This represented about a quarter of the grain normally grown in these countries and a supplement to the depleted local harvest of one-third. For the future, it is reasonable to suppose that, as a result of the establishment of monitoring systems, aid will arrive more rapidly and be distributed more effectively, as the result of experience gained and, therefore, the proportion of drought-induced deaths will be lowered even further. Thus the gains in raising life expectancy won by mass vaccination programs, by improved water supplies and sanitation, and by medical services and superior child nutrition will not be wiped out by the effects of catastrophic years of drought-induced crisis mortality. The world community has proven to be far more successful in minimizing the effects of disasters of short duration than in solving the long-term problem of persistent and perduring poverty. The question which remains is whether unchecked population growth must, of necessity, lead to continuing desertification which, in the long term, must prove as disastrous as constant drought.

In an ecologically fragile environment, where population is already pressing upon resources, the continuance of rapid population growth—where other factors remain unchanged—must lead to deterioration of the habitat. Where those who have made demography the true dismal science are at fault is in their assumption that other things will inevitably remain equal. Most men no longer live in caves and hunt wild animals precisely because other factors do not remain equal: man learned to build shelters when population growth denied him access to readily available caves. Human populations learned to adapt themselves to their environment and to adapt their environment to themselves. Necessity is the mother of invention not only in the developed but also in the developing world. The fatalism of the demographic doomsday men is far greater than that of the much-maligned rural populations of the third world. To anyone travelling through the Sahel during the drought, most impressive of all, even more impressive than the devastation, was the tenacity of the people, their capacity to adapt and change their habits, their diets, their residence, and their very way of life in order to survive.

The Demography of the Sahelian Countries

In order to plan for the future of the region it is necessary to understand the current demographic position.[1] In mid-1974 the six Sahelian countries probably contained some 25 million people unevenly spread over 2 million sq mi of territory, much of which is desert. Population densities are much lower than those to be found to the south in the forest areas of the coast or even in northern Nigeria, a savannah region which contains more than half of the total population of the whole West African savannah. The only available data drawn from the French administrative censuses would suggest that, whereas 50 years ago the population was increasing at an average rate of 1% per annum, since the Second World War it has increased by 3% per annum (Caldwell, 1975b).

This rise in growth rate is the result of a marked decline in mortality, but mortality levels in the Sahelian countries are still very high. When the last French surveys were carried out in the early 1960s the expectation of life at birth was no greater than 35 years, infant mortality was approximately 250 per thousand, and crude death rates were in the upper 30s per thousand. Nomads in the 1970s were probably still experiencing such levels of mortality with crude death rates approaching 40 per thousand even when not affected by the drought. (The only study of the demography of nomads in West Africa reported lower death rates but was bedevilled by underreporting as is shown in the totally implausible sex ratios at birth; in Ganon, 1975.)

Such high death rates set a firm limit to the rate of natural increase. Paradoxically, they also limit the possibility of fertility regulation. Fertility data for the region are more reliable and would suggest birth rates in the upper 40s reaching 50 per thousand or above (Page, 1975). The nomad birth rate is probably somewhat lower (Ganon, 1975) and there is reliable evidence of lower birth rates among the nomads than among the sedentary population in the Sudan (Henin, 1969). Thus a non-drought rate of natural increase for the Sahelian countries would probably be little over 1.5% per annum, while the rate for the northern nomads would be no greater than 1% and could even be lower (Caldwell, 1975). If these estimates are reasonable and if mortality continues to decline gradually while fertility remains constant, then the nomad population of the region would increase by two-thirds by the close of the century, while the total population would treble within the next 50 years. That would mean that the total population of the region would rise by 17 million persons by the close of the century.

It would be unreasonable to suggest that the available demographic data for the region are of high quality. There is a close relationship between levels of development and the quality of demographic and other statistical data, and an additional complicating factor in francophone Africa is the fact that population is correctly associated in the popular mind with tax burdens and, therefore, undercounts are almost inevitable. In Niger, the Faulkinghams found that only after a year's residence in the village were they able to secure accurate census information from the more suspicious members of the community (Faulkingham and Thorbahn, 1975). Their study of a single village provides the only detailed demographic data on the effects of the drought gathered within a longitudinal framework. (Retrospective data gathered for such a troubled period are extremely difficult to obtain or to interpret.) Remarkably, these data show that for this small area with a total population of 1,472 in 1973, farming 720 ha of plateau and valley land, death rates actually fell during the early years of scarcity and drought, while the birth rate fluctuated about a constant level. Thus the crude death rates were 17 per thousand in 1969, 15 in 1970, 19 for 1971, 11 for 1972, and 12 for 1973; birth rates for the same years were 54, 39, 47, 37, and 53. Here the worst pangs of hunger began in January 1974 when 25% of households were eating leaves from the trees at least once a day and many households had fasted for 4 days in 10 preceding the interviews. The death rate for 1974 rose to 43 per thousand, largely as the result of an epidemic of spinal meningitis of such intensity that more than a quarter of all children under five years of age in the village died.[2] At the same time the birth rate dropped to 32 per thousand, possibly as a result of greatly increased foetal mortality in very poorly nourished women.

It is impossible to say how typical these data for a single village are of the Sahelian agricultural zone as a whole. However, they do serve as an interesting commentary upon some features of the situation. People very rarely die of starvation as such. They become weakened by a grossly inadequate diet and then succumb to infectious diseases which, had they been in better condition, they would have survived. Deaths in the very youngest children who were still being breast-fed were less common than among those who had been weaned; the latter group suffered almost double the risk of mortality. The rise in the death rate, when it finally arrived, was of catastrophic proportions and was sufficient to return the population to a no-growth situation. It should be noted that this particular village was blessed with singularly good access to health facilities, being only 20 km from the administrative center of Madoua.

One remarkable feature of the response of the villagers themselves was their extreme adaptability. Given the food shortage, they did everything possible to raise their cash incomes, from handcrafts to growing vegetables on irrigated land patches to migrating. During the period from 1969 to 1974 the proportion of males aged 15 to 44 who left the village to seek seasonal employment elsewhere rose from 35 to 75%. The principal effect of this migration was that there were fewer mouths to be fed at home. The men's remittances home to the village were of less significance. Thus the village is far from being a closed economic system (even the vegetables were sold to itinerant traders). In many ways this may be typical of the Sahelian countries, which survived the drought without major loss of life as much by the flexibility of local reactions as by the aid brought in from the outside by international groups. Although this village was only 25 km north of the Nigerian border, there was very little evidence of the migration of whole families. However, whole families which already had strong kinship links with other groups further south did move from neighboring villages.

Struck by a rate of natural increase that remained at 3% per annum through the early years of the drought, Faulkingham and Thorbahn (1975) concluded their original study with the comment that

> the gravity of the inexorable clash between food supply and population increase cannot be emphasised strongly enough. If Tudu is representative of villages of sedentary farmers throughout the Sahel—and we believe it is—then aid efforts aimed at reducing mortality, dietary deficiencies, and disease should be secondary to the development of coherent strategies to bring food production and distribution policies in line with population increases. Clearly, the drive to reduce mortality will treat only the short-run problem and in a very real way will compound the difficulties the farmers of the Sahel face. It seems to us more feasible to implement this strategy by placing greater emphasis on family planning efforts than on agricultural development.

It is only fair to note that this was written before the village had seen over a quarter of all children under the age of five die within a period of a few months.

There are two vital questions concerning restraint of population growth rates in this ecologically deprived region, one relating to the possibility of such restraint and the other to the desirability of such a goal. In the projections presented above, Caldwell assumed that birth rates within the region would remain constant for the remainder of this century. In support of this assumption he presented data on levels of antinatal knowledge and practice among Sahelian women in 1970. These suggested that even in the capitals and large towns of the Sahelian countries, with the exception of Senegal, fewer than one married woman in a thousand was currently using modern contraception. This would hardly imply a hopeful prognosis for future fertility decline even in the urban areas. Data from Dakar, the one sophisticated metropolitan city of the region, show that in the age group of 20 to 24, 45% of women have knowledge of modern contraception but that only a maximum of 6% of the most protected age group have used either the pill or the intrauterine device, or IUD (Guitton, 1973).

Elsewhere in the world there is considerable evidence that women wish to have fewer children than they are actually having. However, even in relatively sophisticated coastal Africa all the available evidence suggests that the desire to limit family size is confined to a small minority of the educated (Ware, 1975). Three comparable surveys of women in Upper Volta, Niger, and Ghana found that less than 1% of rural women in the two Sahelian countries thought that women should have small families, while in comparatively rich cocoa farming areas of rural Ghana the comparable proportion was 25% (Pool, 1972).

It should be strongly emphasized that, although ignorance of modern contraception is almost universal in the Sahelian countries (Caldwell, 1975, estimates that only 0.5% of rural women know of any modern method of contraception), ignorance is not the major barrier to fertility reduction. The real barrier, which is much more difficult to break, is the fact that parents have no wish to have fewer children. Indeed, many wish to have more surviving children than now remain to them. There is also a strong possibility that rural birth rates will actually rise in some areas as a sequel to the spread of improved health facilities. This is especially likely in those areas of eastern Niger and Chad which form part of the

central African low fertility belt, where fertility levels are abnormally low and where there are very high proportions of women
who have had no children, presumably as a result of medical
factors which are removable (Page, 1975).

It is very difficult for the western observer, seeing swarms of half-
naked children in the midst of a cluster of mud huts, to resist the
belief that their mothers must feel harassed and their fathers must
find them a heavy drain upon the family budget. It requires a real
effort of the western imagination to accept that this is a region
where a man's wealth is measured by the number of his children
and his wives, and where mothers of six children pray to have
more.[3] The traditional remedies of the region do include contra-
ceptives and abortifacients but these are for use in the case of
unsanctioned premarital and adulterous unions, and not to reduce
the total number of one's children. Indeed, throughout the region
the most common complaints brought to the traditional healers
concern sterility and secondary sterility. Thus, the desire for large
families is not simply the result of a fatalistic acceptance of the
inevitable but a deeply rooted positive value within the society.

It is often implied that the rural populations of developing
countries continue to have large families because they are incapa-
ble of making and acting upon the cost-benefit analysis which
would reveal to them the irrationality of their behavior. Yet any
detailed analysis of the economics of the large family in this region
shows that the larger the family the greater the benefit to the
parents. Even Faulkingham and Thorbahn (1975), who advocate
family planning, admit that "the large family makes eminently
good sense. As land is perceived as unlimited in its availability, a
man and woman insure their own security in old age by having as
many children as possible, for the cost of feeding children is more
than offset by the value of their labour in producing food (both in
farming and in the cash sector) in the years ahead."

The large family system is in fact a rational response, and
possibly the only rational response, to conditions in the region.
Where up to a quarter of all children die before their first birthday
what motive can parents have to limit the number of their children? (Even in the seemingly favorable conditions of Tudu, only
80% of children survived to age five.) Children are an investment,
one which provides the parents' only security against sickness and
old age. Even when still young their contribution to the labor force
is considerable, partly because they save the parents from menial
and time-consuming tasks—scaring birds, fetching water, carrying

wood—and partly because their presence permits the greater specialization of labor, allowing, for example, the pastoralist to segregate sheep, cattle, and goats, and find the most appropriate pasture for each. As early as the age of ten, a child's productive power may be as great as his consumption. Even where this is not the case, however, the parents have no alternative strategy by which they can secure the future. It is also true that the man who by chance has few children does not thereby escape the burdens of child rearing. He is merely expected to take a greater share in providing for the children of his siblings.

Even if it were true that the Sahelian countries would benefit from a slowing in the rate of growth of their human populations, it is also true that there is currently no mechanism whereby the advantages accruing to the nation as a whole are reflected in the advantages to the individual family which chooses to limit the number of its children. While it is certainly possible to envisage a system in which the interests of individual families are in conflict with those of the nation as a whole, in this case the apparent conflict should serve to suggest that the national interest may have been improperly defined. Poor as the Sahelian countries are now, they were considerably poorer 50 years ago when their populations were a third of their present levels. The burden of proof must lie with those who argue that food shortages in the Sahel result from population pressure.

Agriculture and Population Growth

Malthus argued that the increase of population to a level beyond the carrying capacity of the land must result in the elimination of the surplus population either by direct starvation or by other positive checks, which are the direct result of insufficient food supplies, such as warfare.

> The new version of Malthusian theory is based on the idea that the increase of population leads to the destruction of the land and that people, in order to avoid starvation, move to other land which is then destroyed in its turn. The neo-Malthusians collect all the evidence on the misuse of land and paint a picture of the world as a place where growing populations are pressing against a food potential which not only is incapable of increase but is

even gradually reduced by the action of these growing populations (Boserup, 1965).

Population carrying capacity studies are inherently Malthusian in conception because of their almost invariable assumption of conditions of constant technology. Thus, during the drought, the U.S. Agency for International Development (AID) commissioned Priscilla Reining to carry out a study to determine responses to the following questions: "Is there population pressure on the land resources and how can you tell with ERTS (Earth Resources Technology Satellites)? If there is population pressure, is this leading to increased desertification especially under conditions of increased climatic aridity?" These questions were then redefined to result in an inquiry into "the actual, the potential, and the optimum carrying capacity or relationship between land and man, for a given group, at a given level of technology, and in a given climatic regime" (Reining, 1973). To calculate carrying capacity for a given climatic regime would appear to be only reasonable, but to assume a constant level of technology is both to prejudge the issue and to indulge in a degree of fatalism more often imputed to ignorant subsistence farmers. Reining's calculations were based upon a formula which derived the area required per capita by multiplying the proportion of the cultivable percentage (land which is, has been, or potentially could be, put into cultivation) relative to a base of 100 by the land-use factor, which is the ratio between the length of time in crop and the length of time in fallow. Thus all these calculations can reveal is whether there is any land that could be cultivated but which is currently never used. The whole question of whether it would be possible to shorten the fallow period without reducing subsequent yields is bypassed by the assumption of constant technology and constant fallow periods as incorporated into Allan's original formula and Reining's adaptation of it (Allan, 1965). Yet Reining observes with some surprise that, as a result of the "manure gradient," it was possible to maintain the fields closest to the village in "apparently continuous use." It could plausibly be argued that the only constraints on more widespread continuous cropping were a lack of labor—this is an area where agriculture with hand-held tools is effectively horticulture—together with a lack of manure due to the limited animal population and to the ignorance of the use of green manure. For example, the Kano close-settled zone supports much greater densities of population on soils which are little superior to those of the villages studied by the use of intensive traditional agricultural methods (Mortimore and Wilson, 1965).

It is typical of the unidimensional framework of such studies that Reining argues that the absence of high proportions of men from the three villages is confirmatory evidence of population pressure upon the land. Men in fact leave frontier areas where land is to be had for the asking to move to the towns simply because the returns to labor are higher in the towns (Donald, 1970). Migration results from a combination of attractive features of the receiving areas and repellent features of the source areas. Even in the lush agricultural areas of the coast, urban incomes are far higher than those of the farmers. Inland, the contrast can be even sharper; upper level government officials in Upper Volta, for example, receive salaries 19 times as great as those of wage-earning rural laborers. The cash incomes of many subsistence farmers would be even lower (Elliott, 1975).

The problem with any measure of land carrying capacity in which technology is held constant is that land is likely to appear to be fully utilized, except at remarkably low densities where vast tracts of land remain unused. Indeed, if land that could be cultivated is lying totally unused, then the alternative question arises as to the nature of constraints which prevent population growth in the absence of any shortage of resources. Once a fallow system is introduced and calculations of carrying capacity assume a constant technology, the existing rate of rotation will always appear to be the maximum possible.

Jacks (1956) described three stages in the development of man's use of the soil. In the first stage ecological balance is maintained because shifting cultivation does not permanently deplete soil fertility. In the second stage soil exhaustion occurs because man effectively mines the fertility of the soil without replacing plant nutrients removed by cultivation or erosion. In the third stage man has learned to conserve the soil by returning as much or more than is removed, so that permanent cultivation of a single area becomes possible. Allan (1965), the originator of carrying capacity studies in Africa, and those who prophesy doom as a result of continued population growth in the Sahel, appear to assume that the aim of agricultural development should be to maintain soil fertility by preserving the balance between loss of fertility through short cultivation periods and replenishment through long fallow periods. More adventurously, Gleave and White (1969) argue that "to forestall permanent loss of fertility, there is a strong case for hastening passage into the conservation stage to preserve land that is capable of supporting greater numbers if a more efficient farming system is practiced. But over much of West Africa, there is a lack of manpower to support such a system."

Given the current tendency to view all population growth as an inevitable evil, it must appear to be extremely paradoxical to argue that greater populations are necessary to ensure more efficient and more conservational agricultural practices in the savannah. Obviously if it is possible to farm the land more efficiently with greater population densities, populations could secure the same benefits by grouping more closely together. Reining gives evidence of soils with a carrying capacity no greater than 12 persons per sq km under existing technology. Certainly more intensive habitation than this would bring considerable social benefits in terms of facilitating the provision of services for education, health, and the like. The question that remains is why existing populations do not adopt more intensive techniques. The answer, as Boserup (1965) has cogently argued, is that intensive agricultural methods require considerably greater labor inputs per unit of output than does shifting cultivation or rotational bush fallowing. After a very detailed study of the farming practices of the Ba Dugu Djoliba of Mali, Jones (1970) concluded

> Djolibans are motivated by pride in being good farmers, by desire to pay their taxes and avoid outside harassment and by desire for the things money can buy. However, they have other uses for their time too, long before the Djoliban farmer would farm to the physical labor maximum—denying himself all time for leisure and social activity—he would seek non-agricultural ways of getting cash.

In this particular case the establishment of a tomato canning plant in the region, one prepared to take unlimited quantities of tomatoes at a fixed price, demonstrated that the Djolibans were willing to farm intensively if they could foresee a sufficient return for their labor.

The tendency to consider irrational those agriculturalists who do not move towards European-style continuous cropping systems stems from a more general belief that the life of primitive peoples must be nasty, brutish, and short. That the existing way of life may be perceived as being more pleasant than the possible alternatives is simply not taken into account. Yet, as Lee (1965) has shown for the Kung Bushmen of the Kalahari Desert, "a carefree and secure existence is possible in a society of hunter-gatherers living in a harsh, semi-desert environment." Even in these conditions, it is possible to secure a sufficiency of food when men and women spend 40 and 25% of their waking hours, respectively, in food procuring activities. The remainder of the Bushmen's time is free for talking, visiting, and resting. When asked for their views on agriculture, the Kung simply reply, "Why should we plant when there are so many mongono nuts in the world" (Lee, 1965).

At each stage of intensity in the exploitation of the environment it is possible to cite examples of peoples who do not wish to move to more intensive forms of exploitation, precisely because such changes would involve more active labor on their part. Thus, in a discussion of "Pakot Resistance to Change," Schneider (1959) explained that the Pakot cattle herders of Kenya were reluctant to grow maize in addition to millet, even having to be forced to make the innovation, because the maize needed more attention and better growing conditions than did the traditional crops. In addition, maize entailed more labor for the women who had to grind it.

Speaking of neighboring groups with a similar resistance to change, Hennings (1951) argued that a lack of change does not necessarily imply stagnation.

> The Elgeyo herdsman has as good food or better, as good or better housing, and a generally more healthy life than millions who live in the industrial towns of Europe. Unemployment and poverty, those spectres of western civilization, have no meaning for him. He is at one with his environment, a pastoralist with his cattle on the high equatorial pastures of Africa . . . he is carefree and happy in a way that millions in Europe never know.

Resistance to agricultural change was studied by Rowena Lawson (1967) in Lower Volta, Ghana. This is an area where agriculture is highly labor-intensive, and where the average value of the nonlabor inputs into agriculture totals less than 5 lbs per year per household (and it is largely confined to the planting of seedlings). The size of the holding is almost entirely a function of labor input as "there is no scarcity of land." The farmer who planted the largest area was the man with the most wives (4) and with the largest supply of children who could join in the work (15). Between

1954 and 1964, the average area of land cultivated per man engaged in agriculture had risen from .55 of an acre to .82, not as the result of any technological innovation but simply because men were working longer hours. Even so, in 1964 farmers only worked an average of 4.1 hours per day on the 174 days of the year on which they went to their farms. In contrast, funerals, travelling, and communal labor accounted for an average of 68 days, while 84 days were spent resting.

> With this low level of labor input there are obviously great opportunities for increased productivity. Labor could work more intensively throughout the year, and more critically, could spend more hours per day on the farm during periods of peak labor input. There is a local demand for food, marketing facilities are good, there is a good road regularly plied with lorries, and yet there appears to be a reluctance to exploit increased demand by increased production (Lawson, 1967).

Such a society is a very long way away from both the ecologists' conventional scenario of population pressing inexorably against resources, and the economists' vision of profit motives resulting in a similar maximization of exploitations levels. Lawson, an economist, comments almost with bitterness that "the unquestioning acceptance of a traditional standard of living is a fundamental impediment to economic growth." Farmers in this area are able to accumulate savings which they invest, not in their farms, but in cattle which have the advantages that they breed their own interest, are looked after by another ethnic group in return for milk, and can be instantly exchanged for cash at any time (Lawson, 1967).

The strongest evidence that less intensive agricultural practices are favored because they allow greater leisure to the farmer comes from those areas where intensive cultivators have reverted to more "primitive" practices. For example, the hill tribes of the Kauru hills of Zaria, Nigeria, gaining access to land on the plains as a result of the colonial peace, abandoned intensive practices of permanent cultivation and terracing for rotational bush fallowing (Gleave and White, 1969). Such changes do not simply reflect a relief from pressure on land resources but also reflect the creation of a superior economy, compared with the old intensive economy (Morgan, 1969), in terms of its returns for labor expended. Extensive methods of shifting cultivation are almost invariably more productive for farmers limited to hand tillage, with no machinery or draft animals. Contrary to popular belief the more "primitive" the agriculture, the less labor it requires.

The apparently academic debate on the direction of the causal relationship between population growth and agricultural innovation is of great practical significance for the savannah agriculture of sub-Saharan Africa. If the Malthusian case is closer to reality, then every effort should be devoted to the attempt to restrain population growth which would represent a forlorn but solitary hope. If, on the contrary, the lack of agrarian change in the Sahelian countries actually reflects the lack of intense population pressure to date, the prognosis is much more hopeful.

Historically, the evidence favors the view of the relative under-population of the Sahelian countries. Before the colonial peace, marauding bands of pastoralists, warriors, or actual armies raiding for cattle and slaves encouraged intensive cultivation around nucleated villages in concentric zones, with the intensity of cultivation rapidly decreasing as distance from the center increased (Savonnet, 1956, 1959). Since the peace, shifting cultivation has advanced into territory which had formerly been the province of the pastoralists. This reflects the choice of a less laborious form of agriculture as much as direct population pressure. Having very little sense of history, those who prophesy doom for the region fail to realize the extent to which many of the phenomena which they interpret as the result of increased population pressure are, in fact, the result of a change in the balance of power between pastoralists and sedentary cultivators. Before the forcible imposition of peace by the colonial powers, the pastoralists were able to enforce militarily the retention of lush riverside lands for pasture. The expansion by the agriculturalists into previously uncultivated areas is now positively encouraged by African governments—in which the nomadic groups have minimal, if any, representation. (Mauritania is a special case having a nomad-influenced government but also great mineral wealth and little scope for settled agriculture.)

Under certain agricultural conditions there are actually minimum densities necessary for successful farming. Thus on the fringes of the tsetse belt it is possible to extend the tsetse-free area by clearing the banks of streams and other areas of scrub vegetation. However, this is only possible where the population reaches densities on the order of 27 persons per sq km. In a less densely settled area there is an insufficient supply of surplus labor for the task (Grove, 1961). There is also some agreement that, as a result of the short duration of the rainy season and the abundance of weed growth associated with it, hand cultivation by one individual cannot be extended much beyond 0.8 ha (Grove, 1961; Hunter, 1972). At a more general level it is also true that the Sahelian pattern of a series of islands of intense agricultural and economic activity separated by vast, almost empty, open spaces greatly increases the costs of providing and maintaining roads and railways. These high costs of transportation, consequent upon low population densities, slow down that transition to a modern money economy without which one cannot have a modern agricultural system (Dumont, 1969).

Human population growth is not the only pressure upon the ecological resources of the savannah agricultural areas. The introduction of the growth of crops such as cotton and groundnuts for the export market has created a situation in which there can be very heavy pressures upon land resources, quite independent of the size of the human population which has to be fed. Originally, the reluctance to devote time and land to such crops was so great that the French colonialists found it necessary to introduce taxes which had to be paid in cash in order to provide an inescapable motivation for moving into the market economy. Cotton is particularly destructive of soil fertility when grown with the technologies and resources currently available in the savannah. This is a crop whose spread is clearly associated with the pressures of modernization rather than the pressures of population growth (Lallemand, 1974). Those who blame the farmers of the region for their alleged thoughtless destruction of natural resources should ponder on the reluctance to grow cotton, which was originally introduced into some areas of Chad at the point of a gun. The dust bowls of America and Australia resulted more from the desire to make money than from the need to feed the populace. In sub-Saharan savannah Africa, destruction of the environment for motives of monetary gain has been largely restricted to European companies, which have been content to "mine" the agricultural resources of the area (N'Dongo, 1976).

Provision for smooth transfer from rotational to permanent cultivation should begin in the vicinity of the towns where pressure upon land is greatest, market agriculture is the rule, and sufficient labor is available. Here the aim would be to introduce green manure crops, the use of artificial fertilizers, and genuine mixed farming. Cultivation in the savannah will always be difficult because "the outstanding characteristic of the rainfall over most of tropical Africa from the agricultural point of view is that, averaged over the year it is less than, and often considerably less than, the amount of water a crop well supplied with water would transpire" (Russell, 1968). In that sense drought is the normal condition. Contrary to popular belief there is very little evidence of available technologies that can quickly raise the productivity of savannah agriculture. There are no semiarid regions of the world from which the savannah farmers could copy methods of agricultural improvement appropriate to their aims and their resources. The examples of what can be done in the United States or in Israel with the resource bases of those countries are totally inappropriate. There is, however, one African precedent: the Kano close-settled zone. Here, intensive agriculture supports densities of up to 500 persons per sq mi in the savannah but only because Kano, the great city with a population now nearing a million, is there to provide the market, the manure, and the dry season employment to support such densities.

There are no simple solutions to the agricultural problems of the region, as one commentator on Upper Volta's agricultural future tersely suggested, except emigration (Hance, 1967). To take but two examples of the problems involved, fertilizers and improved grains, is to appreciate this. In the transition from shifting to permanent cultivation, fertilizers are obviously vital. Unfortunately, however, as FAO's field trials across West Africa have shown, as a general rule, the lower the rainfall the less the effectiveness of fertilizer applications in raising yields of nonirrigated crops. Sorghum and millet, or at least the varieties currently available, are also the least responsive to fertilizers of all the West African staples. Very little has been done to produce improved varieties of these staples. Yet the development of a fast-ripening sorghum that would make ploughing possible between harvest time and the onset of the rainy season would be invaluable, as would varieties made less attractive to birds through the development of extensive sheaths.

It is a mistake to think that because poor-country farming systems use primitive tools and techniques they are therefore simple. . . . The peasant farmer's reputation for conservatism reflects in large measure scientists' frustrations in their attempts to improve peasant farming. In many cases the peasant farmer's reputation is unwarranted; the facts show scientists to be offering correct but irrelevant information (Joy, 1971).

As Lord Lugard (1922) observed half a century ago, "the peasant is more influenced by the success of his neighbour than by any amount of advice." In certain irrigated areas of the savannah one can see the Chinese demonstration rice farms with the irregularly planted rows of the African farmers alongside them. In this case imitation is more than sincere flattery.

An excellent example of correct but irrelevant information relates to the use of fertilizers, which would undoubtedly raise the yields of cereals but not necessarily to the profit of the farmer. Even those most in favor of fertilizers admit that "if any government in tropical Africa should wish to increase the production of cereals by using fertilizers to produce more food for the increasing population, it might be necessary either to stabilize grain prices at a higher level or to subsidize fertilizers or both" (Richardson, 1968). About the maximum return that can be expected from the use of fertilizers in nonirrigated savannah soils is a 50% increase in yields. Yet at the same time, the use of fertilizers also encourages weeds, making the crops more labor-intensive. In addition, the application of nitrogenous fertilizers early in a subsequently low rainfall season can actually reduce yields. Accurate weather forecasting in many cases would do more to raise yields than would the application of fertilizers, under present constraints. However, a central problem in raising the productivity of savannah agriculture remains the lack of a market.

The Nomads

In his study of the relations between pastoralists and settled agriculturalists in the Sahel, Jean Gallais (1972) argues, with considerable historical and anthropological evidence to support his case, that the settled farmers formerly kept a large number of cattle. However, as the Moors, Tuaregs, and Fulani came down from the north into the political vacuum which followed the decline of the Sudan empires of the Middle Ages, they were able to establish a quasi-monopoly of cattle raising. This area of Africa was then in the course of demonetization, and cattle and slaves became the only forms of accumulated wealth that could be extracted as tribute from the vanquished or the "protected." The only farming groups which kept their cattle were either those which produced a military aristocracy, such as the Mossi, the Hausa, and the Serere, or those which fled to the most inaccessible areas, such as the Kapsiki of north Cameroon. The division of labor between pastoralists and cultivators can thus be seen to result from political factors rather than from any ecological necessity. Indeed, close relations between the two are ecologically very advantageous.

The magnetic attraction and repulsion between the pastoralists and the cultivators are reflected in the seasonal rhythms of the relationship between the two groups. The pastoralist returning to his winter camp to clear a field finds that the place, abundantly manured by his herd during the previous year, has attracted the cultivator. The latter sows for preference on the fertile cattle tracks around the watering points, as if deliberately seeking to quarrel with the pastoralist. Conflict is exacerbated as the fields of the cultivator are open, whereas those of the pastoralists are fenced. Hence there are brawls, as herds damage crops. Once the harvest is gathered, a period of amicable relations follows, known by many nomadic groups as "the time of wealth." It is then that barter is stimulated by the abundance of milk and grain and that the flocks feed on the millet straw while manuring the cultivators' fields.

Despite numerous counter-movements, even before the drought, there had been an overall decline in the area devoted to pastoral activities. This reduction in pastoral area was not simply the result of encroachments by cultivators driven to expand the cultivated area by the pressures of population growth. Equally important factors were the introduction of cash crops, notably cotton and groundnuts, which made it worthwhile to cultivate a greater area than that needed to satisfy subsistence needs; the introduction of irrigation schemes which either swamped good pasture lands on the river banks or preempted them for cultivation; and the political changes that have allowed the cultivators to revert to more extensive forms of agriculture in areas where there had formerly been political pressures constraining them within limited regions which could easily be defended.

Still, whatever the cause, this reduction in pastoral area raises the problem of the sedentarization of the nomads. This is the theme of all administrative ideas and, with rare exceptions, all administrative action on behalf of the nomads. A favorite text is Febre's *La Terre et l'Evolution Humaine* (1922) with its theme that "nomadism is not, and cannot be, a permanent condition, a sort of divine

malediction weighing without hope for the sorely tried." Those
who should be most sensitive to this, however, object because the
nomads themselves give no evidence of a distaste for their way of
life. Although there is some evidence of spontaneous sedentarization in the Sahelian countries, it would be an exaggeration to
define this as an overall tendency. It is, for example, not as strong as
the move which, in the nineteenth century, separated three-
quarters of the Fulani from their pastoral way of life. However,
there is evidence of increasing constraints being placed on the
pastoralists' mobility. Agricultural interests are being taken up,
custody of animals is increasingly kept within the family, and the
salt cure at a distance—which comes in the midst of the most labor-
intensive period of cultivation—is increasingly abandoned in favor of the use of the local salt springs or of blocks of salt.

On the same axis of development, however, the sedentary
cultivators are mobilizing and dispersing. The Sahelian villages
are bursting forth wherever water resources permit it, abandoning
the large villages developed for defense for smaller centers that are
less prone to exactions of taxes and forced labor. Temporary straw
structures in the bush may belong either to nomads or to cultivators colonizing new lands for cash crops.

In looking towards the future, it is realistic to envisage the
extension of a style of life based on an agropastoral economy, with
movement between two or three fixed points not far distant from
each other according to season. This pattern is already found
among groups with a double cultural heritage such as the Rimaibe
and the Iklan, the former slaves of the Fulani and of the Tuareg,
respectively. But not all nomads will choose this path. A few,
living in the northern Sahel, will be able to maintain a fully
pastoral way of life. However, there will also be demographic
pressures on some nomads to change their way of life completely
and to move southwards to the towns of the savannah and even to
those of the coast. For a nomad family to have some members in
town, especially when they have secured some education, is an
excellent insurance against disaster, as many found during the
drought. Insurance, in the form of spreading one's resources as
widely as possible, has always been a feature of nomad culture. The
possibility of some finding employment in the towns has merely
increased the range of options available to them.

The advantages of maintaining an equilibrium between the two
types of land exploitation have long been demonstrated. The
agropastoral system of the southern and eastern border of the erg

(sandy desert) of Niafounke enables the region to support a sizable population of cultivators as well as local cattle and numerous nomad flocks. The delimination of and the respect for transhumance tracks, the host family system (in which each pastoral family is virtually "adopted" by a cultivator family in a union symbolized by the handing over of a bucket and rope granting access to the well), the enclosure of cultivated areas, the disciplined observance of seasonal changes, are all factors which help to maintain this balance.

The demographic pressure of the cultivators upon the pastoral areas is strengthened by their greater rate of natural increase. Thus, in Niger the growth rate of the sedentary Iklan is 3.5% per year, of the nomad Iklan 2.3, of the Tuaregs 1.2, and of the Bororo Fulani 1.1% (Gallais, 1972). It would appear that in levels of mortality the nomads are not systematically disadvantaged, but that they have lower levels of fertility (Henin, 1969). Those who advocate the sedentarization of the nomads should bear in mind that one unintended consequence of such a policy would almost certainly be a rise in population growth rates among the ex-nomads.

The effects on the ecology of pastoralism as a way of life are still the subject of much debate. For the best recent discussion of this, see Monod (1975). Swift (1975) argues that the Kel Adrar (of northern Mali) herding strategy of maximizing herd size must in the long run lead to

> a steady degradation of the land and accelerated erosion. The overgrazing around the hot season wells leaves large areas of soil unprotected at the onset of the rains, which arrive in violent storms. Instead of percolating slowly through to recharge the ground water table, the rain runs off in flashfloods in the wadis. Water remaining on the surface is unprotected from sun and wind, and much of it quickly evaporates. The high winds which accompany early rains blow away the unprotected soil. The result is rapid land degradation, which is now going on actively throughout the Adrar.

The Kel Adrar really have very little choice.

> A strategy of limiting herds to the carrying capacity of the environment would demand voluntary human population stabilization; such a strategy would be extremely

difficult to operate and would give no advantage in bad
years. It is also made difficult by the divergence between
rational individual behaviour and rational collective
behaviour in the use of common property resources such
as pasture and water. Kel Adrar herding strategy is
dominated instead by the fact that animals are mainly a
capital stock, by the backward-bending supply curve in
livestock arising from the fact that the Kel Adrar sell to
raise a fixed target income, and by the need to fuel the
system of loans, exchanges and gifts of animals which
gives social insurance to the individual herdsman in case
of disaster (Swift, 1975).

Any system which provided the maximum protection in bad times,
rather than attempting to maximize advantages in the good times,
would suffer the same ecological disadvantage of equating
numbers with strength.

   Two factors that were actually intended to help them have served
to worsen the environmental situation of such groups as the Kel
Adrar. One was the spread of improved veterinary medicine and
animal disease controls and the other was the provision of an
increasing number of wells. Although epidemics of animal diseases have not received much public attention in recent years, one
could make a good case for the Sahelian countries that such
epidemics have played as important a role as droughts in contain-
ing livestock populations in the region. For example, the great
cattle plagues of 1891, 1915-17, and 1919-20 are still remembered
with horror across Sahelian Africa. The provision of wells, without a parallel provision for additional pasture facilities, only serves
to aggravate ecological problems. One effect of colonial peace
making was to destroy the political authority of the chiefs who had
the power to enforce conservation measures with respect to pastures. The deposed chiefs of northern Upper Volta could decree
that after intensive grazing certain wells or watering points should
not be used until the pastures associated with them had had time to
regenerate. The decline of their authority is visibly reflected in the
landscape where, as a direct result of the break-down of these
controls, some of the most nutritious grasses and herbage have
disappeared. The colonial government in Kenya tried to impose a
system of restricted grazing areas upon the Pakot, who already had
"an indigenous plan whereby areas of termite-resisting grass in the
neighborhoods were closed for grazing except during droughts and

in the dry season of the year, and fines were imposed on those who trespassed" (Schneider, 1959). Both the case of veterinary medicine and that of the provision of wells are examples of the same truth: the more fragile the environment the more difficult it is to introduce one change without setting off a chain reaction of other changes. The provision of wells apparently solved the water problem only to create new pasture problems and to result in a growth of animal populations greater than could possibly be supported in the environment except in the years of the very best rainfall conditions. The growth of herds has always been a cyclical phenomenon, with herd size peaking in years of good rainfall only to fall back again in years of drought. The provision of wells without the provision of additional pastures only accentuates these cycles of peaks and troughs.

Defenders of the nomads from charges of ecological destructiveness are not wanting. Thus Toupet (1975) argues:

> The nomad has often been accused of provoking desertification by allowing his flocks to make havoc of the vegetative soil cover. We would like to show, taking our examples from central Mauritania, that this accusation put forward by the sedentary farmers, is not invariably founded in fact and that precisely where ecological conditions are most fragile, the nomadic pastoralist, through his way of life and his techniques of exploitation, far from damaging the ecological balance, can contribute effectively to the conservation of nature.

An essential feature of the way of life of nomadic pastoralists is that the pressure of man and beast upon the environment is neither ubiquitous nor permanent. Nomads have been able to survive over thousands of years because they have learned to maintain a balance with nature.

The cleavage between those who accuse the nomads of environmental destruction and those who consider them to be conservers of nature often follows the linguistic frontier. In support of the francophones, it might perhaps be argued that in West Africa all the Sahelian countries fall within the francophone sphere of influence and, until very recently, nearly all of those who have actually lived with the nomads have been francophones. Irrespective of the objective situation, anglophones are much more inclined to see overpopulation than francophones who, to the contrary, have often argued that the Sahelian countries are actually gravely *under*populated (Pradervand, 1972).

Heavy grazing, due to the increased herds in good years for which there is insufficient pasture in less favorable years, has always been a problem. Recently, however, the problem has been considerably aggravated by a very marked rise in the number of livestock, partly in response to their increased market value in the south. Not only the nomads are at fault. In some areas small herds of cattle belonging to sedentary cultivators are a new feature that can lead to localized overgrazing. In Mauritania, Toupet argues, the chief malefactors are the urban cattle owners who do not have the same interest in conservation as the nomads for whom pastoralism is the only means of livelihood (1975). Even in Lower Volta, Lawson (1967) found that agriculturalists were converting their savings into cattle.

Woodcutting is another environmental hazard for which the nomads do not bear the major responsibility. Pastoralists in areas such as the Tibesti cut foliage to feed the camels and to enclose the goats. Lugard (1922) attributed the desiccation of the Sudan to such practices and forbade the indiscriminate felling of trees for fodder in Nigeria. Of greater modern significance is the demand for wood in the towns which, in combination with the firing of charcoal, is denuding the surrounding countryside. The provision of alternative fuel sources and building materials within the reach of the poor in the urban areas is one problem which ecologists, eager to halt the advance of desertification, could well turn their minds to; not that the problem is an easy one to resolve in areas without access to coal or (to date) oil reserves. In the Zinder region of Niger, the Department of Water Resources and Forests plants 100 seedlings for every tree that survives. Less than one-quarter of one percent of the total area of the region is currently afforested. It is estimated that, if a third of the area were covered, this would secure an equilibrium and allow the continued usage of wood for fuel (Niger, 1973). It is noteworthy that in some areas it has actually been found that the destruction of woodlands leads to significant rises in the water table, presumably because the transpiration of water from a depth, raised by the tree roots, is prevented (Carter and Barber, 1956).

Cultivation in marginal areas during periods of high rainfall is one cause of desertification against which it may well be necessary to take preventive action. However, the motivations for such plantings, whether by pastoralists or by cultivators, should be investigated together with possible alternatives *before* any attempt is made to repress them.

The measurement of desertification through the study of vegetation is time consuming and expensive, as on-the-ground surveys are still necessary despite advances in satellite and aerial photography. Clos-Arceduc (1956) used aerial photographs to supplement ground observation in his intensive study of the vegetation zones in Niger. His conclusion was that the zones of the Niamey region had shifted 150 km to the south during the past two centuries—in marked contrast to Lester Brown's 30 mi per year desert advance, which was apparently based on travellers' tales. Unfortunately, there have been no equally rigorous studies of subsequent developments.

Given that there is some degree of desertification on the southern fringes of the Sahara, should this be a matter for grave concern? To the conservationists such a question is an anathema but is nonetheless a rational query. With sufficient effort and a bottomless purse it would be possible to do something to reverse the trend by the planting of linear forests, the creation of lakes by the diversion of rivers, and well-timed cloud seedings. However, all that would be gained would be the conservation of lands which can never be other than extremely marginal. To prevent the current minor infringements of the pastoralists would be impossible without their cooperation. A much more rational policy would be to concentrate all available resources on those relatively small and relatively few areas on the desert fringes which are already favored by nature and communication patterns. Emphasis could then be concentrated upon positive possibilities, such as the provision of alternative sources of fuel and building materials, other than native wood, the supply of young fruit trees, pasture improvement, and the like (Grove, 1973).

The above brief discussion of some of the factors leading to desertification should suggest that population growth is not necessarily the chief culprit. Grazing pressures result as much from commercialization as from population pressure; for example, woodcutting results principally from urbanization; also, the taking up of marginal lands may be the result of a desire to minimize labor, or to spread cash crops, or a result of localized population pressures. Other factors, such as the effects of deliberate bush fires, are totally unrelated to the size of the human population.

The question remains as to what could be done for the nomads, given that the area which they currently utilize cannot support vastly greater populations than at present, and is, in any case, under considerable pressure from settled cultivators. Perhaps the first thing to be noted is that nomads are doing a great deal for themselves, utilizing that adaptability which has secured their survival to date. Even in the field of population control, nomad growth rates are by African, and even by world, standards remarkably low (Podlewski, 1975).

Mary Douglas (1966) discusses the case of the Rendille camel herders of the Kenya highlands who rigidly control the rate of population growth by maintaining a high marriage age for women, marrying off their surplus women to neighboring tribes, and practicing the selective infanticide of males born on moonless Wednesdays and after their brothers have been circumcised. Spencer (1973), in a more intensive study of symbiosis and growth among the Rendille and the neighboring Samburu, describes how the Rendille are a group who maintain a relatively constant size in sharp contrast to the cattle-keeping Samburu, whose population is rapidly increasing. He argues that these rates reflect the rates of growth of the two economies (cattle herds multiply much more rapidly than camel herds). The Rendille manage to keep in balance with their harsh environment both by conscious population control and because some of their members voluntarily choose to become Samburu, sharing their less harsh environment. Perhaps, if there were more studies directly concerned with nomad economies, more would be known of the means whereby they have managed to maintain a balance with available resources. One obvious mechanism is through the bride price, payable in cattle. If a man cannot marry unless he can transfer a set number of cattle, then in hard times marriage rates, and thus birth rates, will decline appreciably.

> Marriage involves the giving of cattle to all the relatives of the bride. And a man wants to marry a girl with as many kinsmen as he can afford so that he will have as large a family as possible. Since there is no effective government protection of the individual, the only satisfaction a man can get if his property is stolen or if his rights are abused is to get his kinsmen to back him up. During times of shortage, or if his herds are decimated by enemy raiding or disease, he can also beg from his kinsmen. The more kinsmen a man has, the more secure his position. A man therefore keeps all his animals, even culls, because he

wants large numbers to give in marriage, rather than limiting his herd to good milking animals (Dyson-Hudson, 1970).

Thus, the Karimojong of Uganda live within a system which places a very strong emphasis upon the maximization of sheer numbers. To date the system has worked well, but the eradication of human and animal epidemics has apparently brought population growth to the maximum which the system can support. The result is that the Karimojong are turning increasingly to agriculture by their own choice without the need for official "sedentarization" programs. According to Scudder (1971), nearly all nomads are dependent upon agriculture to the extent that they eat grain—apart from the Masai there are very few nomad groups who do not. Conversely, agricultural peoples may well revert to being hunter-gatherers in times of crop failure.

There have been very few studies of nomad adaptation to urban life, especially in the Sahelian countries. Bernus's (1969) study of Niamey is one notable exception. As might be anticipated, she found that the groups most receptive to, and indeed most driven to, change were those which were already marginal to true nomad society. Thus it is the former slaves of the Tuaregs, the Bella, and not the Tuaregs themselves, who have most successfully adapted to life in Niamey. The nomads who come to Niamey, having no capital, cannot start as traders and usually find work first as porters or as grain processors for the affluent women in the town. Employment as laborers on a regular wage basis, even where the wage is lower—which it often is—than the sums that can be obtained from such casual work in the informal sector is regarded as a promotion and is much sought after. Longer-established Bella have become skilled workers, such as masons, and even affluent shopkeepers. There is virtually no industry in Niamey and thus no scope for this group to acquire industrial skills to which they would be especially suited both because of their preference for wage labor and their culturally induced readiness to learn new skills. (A Bella would be ashamed not to be able to perform any task which other groups can.)

The arguments for providing urban employment for the displaced surplus nomad population, rather than settling them to cultivate the land, are little considered but are certainly important. There is a great deal of evidence that this is the course that the nomads themselves would prefer. Even before the drought there was a considerable flow of nomads seeking work in the coastal towns of Ghana and the Ivory Coast and in the centers of northern Nigeria. Urban employment offers the great advantage that, owing to the vast differential in returns to labor favoring the urban centers, it is possible to maintain strong links with relatives still in the traditional areas and even to alternate between the two ways of life. Within the traditional framework it is, indeed, much easier to be an absentee cattle raiser than an absentee landholder because the nomads have long accepted looking after the cattle of others as well as their own, whereas communally owned land is only shared out among those who are present to work it.

Without making undue claims for the existence of ethnic mentalities, it can nonetheless be argued that nomads have given evidence of remarkable adaptability; witness, for example, those who work in association with Algeria's desert oil wells. If a proportion of nomads will be obliged by population pressure to adopt a new life style, it appears much more reasonable to help them move directly into the twentieth century rather than to route them via the Middle Ages of peasant agriculture. "Why . . . should the nomad of the arid lands be expected to enjoy becoming a sedentary peasant farmer, or worse still, a peasant farmer isolated in the midst of irrigated fields demanding constant attention?" (Stamp, 1961). The successful practice of agriculture under the conditions of the Sahel requires a skilled training, begun in early childhood. It is not as if the Sahelian countries were blessed with a wealth of good, arable land waiting to be opened up. The aim should be to increase the productivity of those who are already working on the land and not to increase the number of those who are dependent upon the land. This is the more true as a major contributing factor to the failure to improve agricultural productivity in the region is the lack of effective demand for agricultural produce, which development of the towns would provide.

The effect of human populations upon the environment is determined as much by distributional factors as by absolute numbers. More than a decade ago, Dudley Stamp pointed out that the intensification of development of the arid lands with the bringing of water from surface or underground sources to make possible the growth of crops or of fodder for domestic animals was ignoring one of the clearest lessons to be drawn from the history of land use. With the exception of Australia

> the great developments in arid lands in the past have been
> associated not only frequently but usually with cities.
> Civilization in the deserts has often been essentially an
> urban civilization. This is true not only of Nineveh and
> Babylon and of the Indus plain; it is true of the Roman
> *limes* in North Africa, [and] to a larger extent of the
> Mandingo and other empires of the Sahara (Stamp, 1961).

In the Middle Ages the Sahel justifiably gloried in great cities of a
magnitude and sophistication almost unknown throughout other
parts of sub-Saharan Africa. Timbuktu, Djenne, and Gao were
great trading centers with populations of 50,000 persons or more. It
is unfortunate that more is not known of the agriculture which
supported these populations, except that they were heavily dependent upon slaves and the tribute of subject peoples (Bovill,
1933). Admittedly the trans-Saharan trade upon which they depended, long in decline, has now virtually vanished, although it
will be extremely interesting to observe the effect of the completion
of the trans-Saharan highway in this context.

One common feature of the medium- and long-term plans made
for the development of the Sahelian countries as a sequel to the
drought is their emphasis upon the rural sector (CILSS, 1974).
Even the broadest plans rarely extend beyond agricultural and
rural development; development of natural resources; and transport, communications, and manpower needs (ECA, 1973). Thus
the adapted structure of the West African United Nations Development Advisory Team provides for three experts in agriculture,
two each in rural development, hydrology, and economics, together with one transport economist and one mining engineer. There is
apparently little perception of any need to take the possibility of
industrial development into account. Yet, from the point of view of
rationalizing the economy of the region and preventing further
desertification, the need is for intensive agriculture geared to an
urban market, not to the continuation of an improved form of
subsistence agriculture.

Establishing a firm economic base for the inland towns of the
Sahelian countries will present many difficulties. There are few
industries which can logically be promoted, especially with the
current lack of skilled manpower (perhaps groundnut and cotton
processing, textile milling, and tanning are the most obvious

possibilities), and the industries of any one Sahelian country have to compete with those of all the others operating under very similar conditions and catering to equally small markets. Possibilities for the future include meat processing; the construction of cheap carts to be drawn by the cattle of the mixed farms of the future and of agricultural implements in general; and the processing of grains which would save the women of the region from hundreds of hours of repetitive labor each year. Short of further mineral finds, which may well occur (ECA, 1973), the chief hope for the Sahelian countries lies in the export of their cattle and their labor to the better-endowed countries along the monsoonal coast. Here, the nomads of the Sahel have a central role to play, both because of their position as cattle raisers and because they are under the strongest pressure to migrate southward.

Conclusion

Concern with the relationship between population growth and desertification is not new. In the sixth century B.C., Napata, the capital of the Kingdom of Kush near the Fourth Cataract, was abandoned because overgrazing by the herds associated with the capital had resulted in erosion and in the advance of the desert (Dixon, 1972). More recently, Lord Lugard (1922) reported how he had seen the advance of the Sahara to Sokoto, Nigeria. He cited the Resident's account of "the drying up of rivers, streams and lakes, the diminishing rainfall, the dying out of large forests and their replacement by poor scrub, the increase in area where crops will not grow, and the gradual decline in yield on the best watered parts; lastly the never-ceasing movement of races, tribes, families and individuals from north to south." Even after a series of good wet seasons, a first encounter with the Sahel and the northern savannah in the dry season is always a chastening experience: how can people live, and even multiply, in such hostile territory?

Contemporary evidence of the overpopulation of the Sahelian countries is extremely difficult to interpret. One major famine in 60 years does not demonstrate that an area is overpopulated in any real sense any more than a plague of locusts or a war which prevented the planting of crops would. It does, however, point to the need for improved storage facilities and communications to assure the supply of food in such periods. In the absence of such facilities, "the carrying capacity of the land in men and animals is that of the poorest years, not that of the average years" (Stamp, 1961).

It must be remembered that it was not until the latter half of the eighteenth century in England and France, and until the nineteenth century in Scandinavia, that a succession of bad harvests was no longer the invariable signal for a sharp rise in mortality (Habakkuk, 1953; Utterström, 1965). This does not show that before these periods Europe was overpopulated, but rather that markets, transport, and welfare systems were insufficiently developed to cope with natural disasters. Fewer people died in the 1970s in the Sahel than in the 1910s, in part because of international aid which came in from outside, but also because of the improved infrastructure within the region itself, which made it possible to move grain from areas where there was a surplus to the affected zones. One feature of rainfall patterns within the region is that there are very wide variations in levels both over time and over space. Even in the worst years of the drought there were some areas which received a sufficiency of rain timed so that a good crop was possible. (The timing of rainfall in this area is as important to agriculture as the overall amount. Several false starts to the rainy season can be disastrous because they involve a great waste of seed grain and of human resources.)

Any consideration of probable future trends in and provisions for the development of the region, or indeed of plans for the prevention of environmental deterioration in the area, must take into account the fact that the people of the Sahelian countries do not themselves perceive the region to be overpopulated. The majority of the elites of these countries, and of the expatriate francophone development personnel who have worked in the region, would argue that rapid population growth is in itself a precondition of development; that the traditional societies of Africa were technologically stagnant precisely because there was no population growth to stimulate technological innovation and entrepreneurial risk-taking. Wilkinson (1973) has recently made a very strong case for rapid population growth and pressure against resources as a precondition for all industrial development. Furthermore, they would argue that, even seen as a single economic unit, which they are not, the Sahelian countries simply do not possess a sufficient population base for economic take-off. According to this group, the problem is less the shortage of capital than the limited size of the market (EEC, 1967). They claim that larger populations and the ensuing greater densities and economies of scale would produce benefits far outweighing the disadvantages of rapid population growth. Economic unions can create enlarged markets but they cannot resolve the problems of low population density, nor can they stimulate the "challenge" of population growth, the invigorating clash of generations, the receptiveness to new ideas, and the active search for new solutions to be found in a fast-expanding population (Pradervand, 1972).

The most obvious needs for the future planning of the Sahelian countries are three: firstly, that planning should consider the economy as a whole and not consider simply the rural sector; secondly, that planning should be framed not simply in terms of projects suitable for a poor, arid country but also in terms of the cultures concerned; and thirdly, that there should be a recognition that significant fertility reduction, even if desirable, cannot be achieved until the overall level of development in terms of health, education, and urbanization has been raised considerably. For the future, an immediate slowing down of rates of population growth is not an available option. The fight against poverty in the region must have the first priority whether the eventual aim is simply to ameliorate conditions for the people, to prevent further desertification, or to slow down the rate of population growth.

Acknowledgment

This paper has drawn upon the ideas of many different people to whom a listing among the references can only be a very inadequate acknowledgment of my indebtedness to their stimulating qualities.

Notes

1. This paper follows the U.N. convention in referring to Senegal, Mauritania, Upper Volta, Mali, Niger, and Chad as the Sahelian countries despite the fact that only their northern reaches are in the true Sahel.
2. Surveys in such conditions of extreme deprivation must inevitably raise ethical issues of the responsibility of the researcher to record or to intervene.

3. Even in comparatively rich and sophisticated western Nigeria, more than half of all women with six living children positively wish for more (Ware, 1975b).

References

Abercrombie, K., 1967. "The transition from subsistence to market agriculture in Africa South of the Sahara," in E. Whetham and J. Currie, eds., *Readings in the Applied Economics of Africa*, Cambridge, Vol. I.

Allan, W., 1965. *The African Husbandman*, London.

Amin, S., 1974. *Modern Migrations in Western Africa*, Studies presented and discussed at the 11th International African Seminar, Dakar, 1972, International African Institute, London.

Bernus, S., 1969. *Particularismes Ethniques en Milieu Urbain: L'Exemple de Niamey*, Université de Paris, Musée de l'Homme, Paris.

Boserup, E., 1965. *The Conditions of Agricultural Growth: The Economics of Agrarian Change under Population Pressure*, Chicago.

Bovill, E., 1933. *Caravans of the Old Sahara, An Introduction to the History of the Western Sudan*, International Institute of African Languages and Culture, London.

Brown, L., 1974. *World Population and Food Supplies: Looking Ahead*, Conference Background Paper, U.N. World Population Conference, Bucharest, 19-30 August 1974.

Caldwell, J., 1975a. *The Sahelian Drought and its Demographic Implications*, Overseas Liaison Committee Paper 8, American Council on Education, Washington.

Caldwell, J., ed., 1975b. *Population Growth and Socioeconomic Change in West Africa*, Population Council, New York.

Carr-Saunders, A., 1936. *World Population: Past Growth and Present Trends*, Royal Institute of International Affairs, London.

Carter, J., and Barber, W., 1956. "The rise of the water-table in parts of Potiskum Division, Bornu Province," *Records of the Geological Survey of Nigeria*, pp. 5-10.

Centre for Disease Control, 1973. *Nutritional Surveillance in Drought Affected Areas of West Africa (Mali, Mauritania, Niger, Upper Volta), August-September 1973*, U.S. Public

Health Service, Atlanta, Georgia.

CILSS, 1973. Permanent Inter-State Committee for the Fight against Drought in the Sahel, *Summary Report of the Work of the Meeting of the Ministers of the Six Sahelian Countries Affected by the Drought, 7-10 September, 1973,* Ouagadougou.

CILSS, 1974. Permanent Inter-State Committee for the Fight against Drought in the Sahel, *Project Inventory of the West African Countries Stricken by Drought,* Ouagadougou.

Clos-Arceduc, M., 1956. "Etudes sur photographies sériennes d'une formation végétale sahélienne: la brousse tigrée," *Bulletin de l'IFAN,* 18: 677-684.

Dalby, D., and R. J. Harrison Church, eds., 1973. *Drought in Africa,* Report of the 1973 Symposium, Centre for African Studies, School of Oriental and African Studies, University of London, London.

Dimbleby, G., 1971. "The impact of early man on his environment," in P. Cox and J. Peel, eds., *Population and Pollution,* Proceedings of the Eighth Annual Symposium of the Eugenics Society, London.

Dixon, D., 1972. "Population, pollution and health in Ancient Egypt," in P. Cox and J. Peel, eds., *Population and Pollution,* Proceedings of the Eighth Annual Symposium of the Eugenics Society, London.

Donald, L., 1970. "Food production by the Yalunka household, Sierra Leone," in P. McLoughlin, ed., *African Food Production Systems, Cases and Theory,* Baltimore.

Douglas, M., 1966. "Population control in primitive groups," *British Journal of Sociology,* 17: 263-273.

Dumont, R., 1969. *False Start in Africa,* New York.

Dyson-Hudson, N., and N. Rand, 1970. "The food production system of a semi-nomadic society: the Karimojong, Uganda," in P. McLoughlin, ed., *African Food Production Systems, Cases and Theory,* Baltimore.

ECA, 1973. *E.C.A. On-going Activities and Proposed Role in Medium and Long-Term Programmes in the Drought Affected Countries of the Sudano-Sahel,* Addis Ababa.

EEC, 1967. *Etude sur les Possibilités d'Industrialisation dans les Etats Africaines et Malagaches Associés,* Brussels.

Elliott, C., 1975. *Patterns of Poverty in the Third World: A Study of Social and Economic Stratification,* New York.

Faulkingham, R., and P. Thorbahn, 1975. "Population dynamics and drought: a village in Niger," Population Studies, 29(3): 463-478.

Febre, L., 1922. *La Terre et l'Evolution Humaine: Introduction Geographique à l'Histoire*, Paris.

Gallais, J., 1972. "Essai sur la Situation Actuelle des Relations entre pasteurs et paysans dans le Sahel Ouest-Africain," in *Etudes de Géographie Tropicale Offertes à Pierre Gourou*, Paris.

Ganon, M., 1975. "The nomads of Niger," in J. Caldwell, ed., *Population Growth and Socioeconomic Change in West Africa*, Population Council, New York.

Gleave, M., and H. White, 1969. "Population density and agricultural systems in West Africa," in M. Thomas and G. Whittington, eds., *Environment and Land Use in Africa*, London.

Gourou, P., 1947. *Les Pays Tropicaux*, Paris.

Graham, A., 1969. "Man-water relations in the east central Sudan," in M. Thomas and G. Whittington, eds., *Environment and Land Use in Africa*, London.

Grove, A. T., 1961. "Population densities and agriculture in Northern Nigeria," in K. Barbour and R. Prothero, eds., *Essays on African Population*, London.

Grove, A. T., 1973. "Desertification in the African Environment," in D. Dalby and R. J. Harrison Church, eds., *Drought in Africa*, Report of the 1973 Symposium, Centre for African Studies, University of London, London.

Guitton, C., 1973. Etude de la Fécondité à Dakar: Premiers Résultats, ORSTOM, Dakar.

Habakkuk, H., 1953. "English population in the eighteenth century," *Economic History Review*, 6: 117-133.

Hance, W., 1967. *African Economic Development*, London.

Haswell, M., 1963. The Changing Pattern of Economic Activity in a Gambia Village: HMSO, London.

Henin, R., 1969. "The patterns and causes of fertility differentials in the Sudan, with reference to nomadic and settled populations," *Population Studies*, 23: 171-198.

Hennings, R., 1951. *African Morning: Experiences as a District Officer in Kenya*, London.

Hunter, J., 1967. "Population pressure in part of the West African Savanna, a study of Nangodi, Northeast Ghana." *Annals of the Association of American Geographers*, 57: 101-114.

Hunter, J., 1972. "Population pressure in part of the West African Savanna: a study of Nangodi, Northeast Ghana," in R.

Prothero, ed., *People and Land in Africa South of the Sahara: Readings in Social Geography*, London.

Hutton, C., 1973. *Reluctant Farmers? A Study of Unemployment and Planned Rural Development in Uganda*, Kampala.

Jacks, G., 1956. "The influence of man on soil fertility," *Advancement of Science*, 13: 137-145.

Jones, S. W., 1970. "The food economy of the Ba Dugu Djoliba Mali," in P. McLoughlin, ed., *African Food Production Systems, Cases and Theory*, Baltimore.

Joy, L., 1971. "Strategy for agricultural development," in D. Seers and L. Joy, eds., *Development in a Divided World*, London.

Lallemand, S., 1974. "A Yatenga village in the course of the 1973 drought," personal communication.

Lamb, H., 1973. "Some comments on atmospheric pressure variations in the Northern Hemisphere," in D. Dalby and R. J. Harrison Church, eds., *Drought in Africa*, Report of the 1973 Symposium, Centre for African Studies, University of London, London.

Lawson, R., 1967. "Innovation and growth in traditional agriculture of the Lower Volta, Ghana," *Journal of Development Studies*, 4: 138-149.

Lee, R., 1965. *Subsistence Ecology of Kung Bushmen*. Ph.D. thesis, University of California, Berkeley.

Lugard, F., 1922. *The Dual Mandate in British Tropical Africa*, Edinburgh.

Monod, T., 1975. *Pastoralism in Tropical Africa*, Studies presented and discussed at the 13th International African Seminar, Niamey, December 1972, International African Institute, London.

Morgan, W., 1969. "The zoning of land use around rural settlements in tropical Africa," in M. Thomas and G. Whittington, eds., *Environment and Land Use in Africa*, London.

Mortimore, M., and J. Wilson, 1965. Land and People in the Kano Close-Settled Zone, Report to the Greater Kano Planning Authority, Zaria.

Moss, P., and K. Swindell, 1975. "Relations with the environment," in P. Moss and R. Rathbone, eds., *The Population Factor in African Development*, Proceedings of a Conference organised by the African Studies Association of the United Kingdom, September 1972, London.

Neumark, S., 1972. "Economic development and economic incentives," in J. Uppal and L. Salkever, eds., *African Problems in*

Economic Development, New York.

Niger, 1973. "Damagaram," *Niger*, Vol. 20.

N'Dongo, S., 1976. *"Coopération" et Néo-Colonialisme*, Paris.

Owen, D., 1973. *Man's Environmental Predicament: An Introduction to Human Ecology in Tropical Africa*, London.

Oxby, C., 1975. *Pastoral Nomads and Development: A Select Annotated Bibliography with Special Reference to the Sahel*, London.

Page, H., 1972. "Fertility and Child Mortality South of the Sahara," in S. Ominde and C. Ejiogu, eds., *Population Growth and Economic Development in Africa*, Population Council, London.

Page, H., 1975. "Fertility levels: patterns and trends," in J. Caldwell, ed., *Population Growth and Socioeconomic Change in West Africa*, New York.

Podlewski, A., 1975. "Bilan d l'état des connaissances démographiques concernant les écosystèmes pâtures et forestiers des régions tropicales (Afrique)," *Cahiers ORSTOM Série Sciences Humaines*, 12(4): 379-400.

Pool, J., 1972. "A cross-comparative study of aspects of conjugal behaviour among women of three West African countries," *Canadian Journal of African Studies*, 6: 233-259.

Poursin, G., 1974. "A propos des oscillations climatiques: la sécheresse au Sahel," *Annales*, 29: 640-647.

Pradervand, P., 1972. "Population et développement," *Développement et Civilizations*, pp. 47-48.

Reining, P., 1973. *Utilization of ERTS-1 Imagery in Cultivation and Settlement Sites Identification and Carrying Capacity Estimates in Upper Volta and Niger*, final report to AID office, Washington.

Roche, M., 1973. "Note sur la sécheresse actuelle en Afrique de l'Ouest," in D. Dalby and R. J. Harrison Church, eds., *Drought in Africa*, Report of the 1973 Symposium, Centre for African Studies, University of London.

Richardson, H., 1968. "The use of fertilizers," in P. Moss, ed., *The Soil Resources of Tropical Africa*, Cambridge.

Russell, W., 1968. "Some agricultural problems of semi-arid areas," in P. Moss, ed., *The Soil Resources of Tropical Africa*, Cambridge.

Savonnet, G., 1956. "Système d'occupation du sol dans l'ouest de la Haute Volta," in *Symposium de Géographie*, IFAN, Dakar.

Savonnet, G., 1959. "Un systeme de culture perfectionée, pratiqué par les Bwaba-Bobo-Oulé de la région de Houndé (Haute Volta)," *Bulletin de l'IFAN*, 21: 425-458.

Schneider, H., 1959. "Pakot resistance to change," in W. Bascom and M. Herskovits, eds., *Continuity and Change in African Cultures*, Chicago.

Schove, D. J., 1973. "African droughts and weather history," in D. Dalby and R. J. Harrison Church, eds., *Drought in Africa*, Report of the 1973 Symposium, Centre for African Studies, University of London, London.

Scudder, T., 1971. *Gathering among African Woodland Savannah Cultivators, A Case Study: The Gwembe Tonga*, Zambian papers, 5, University of Zambia.

Shorter, A., 1974. *East African Societies*, London.

Skinner, E., 1960. "Labour migration and its relationship to socio-cultural change in Mossi Society," *Africa*, 30: 375-401.

Spencer, P., 1973. *Nomads in Alliance Symbiosis and Growth among the Rendille and Samburu of Kenya*, London.

Stamp, D., 1969. "Some conclusions," in *A History of Land Use in Arid Regions*, UNESCO, Paris.

Swift, J., 1973. "Disaster and a Sahelian Nomad Economy," in D. Dalby and R. J. Harrison Church, eds., *Drought in Africa*, Report of the 1973 Symposium, Centre for African Studies, University of London, London.

Swift, J., 1975. "Pastoral nomadism as a form of land use: the Twareg of the Adrar n Iforas," in T. Monod, ed., *Pastoralism in Tropical Africa*, International African Institute, London.

Toupet, C., 1975. "Le nomade, conservateur de la nature? L'exemple de la Mauritanie centrale," in T. Monod, ed., *Pastoralism in Tropical Africa*, International African Institute, London.

Utterström, G., 1965. "Two essays on population in eighteenth-century Scandinavia," in D. Glass and D. Eversley, *Population in History, Essays in Historical Demography*, London.

Ware, H., 1975. *The Sahelian Drought: Some Thoughts on the Future*, United Nations ST/SSO/33.

Ware, H., 1975. "The limits of acceptable family size—Western Nigeria," *Journal of Bio-Social Science*, 7(3): 272-296.

Wilkinson, R., 1973. *Poverty and Progress: An Ecological Model of Economic Development*, London.

9

Principal Problems of Desert Land Reclamation in the U.S.S.R.

A. G. Babayev

The study of the reclamation of the world's deserts has aroused the deepest and most comprehensive interest, due to rapidly growing populations and to the development of science and technology, which makes it possible to spread man's economic activities on a large scale over lands that to date have been sparsely populated and inadequately used.

Deserts cover about a quarter of the world's dry land with their area in the Soviet Union amounting to about 300 million ha (2.47 acres per ha), mostly within Kazakhstan, Uzbekistan, and Turkmenistan. They stretch over 3,200 km from west to east and over 2,500 km from north to south (see figure 9.1).

Since ancient times, economic activities of the people living in central Asia and in southern Kazakhstan have been based on the utilization of natural resources existing in the deserts. For hundreds of years, however, the reclamation of desert land has been primitive and slow. It was not infrequent that man proved to be powerless to combat the desert and that he waged an unequal struggle which ended in retreat. Nowadays when one travels through the deserts, one encounters many old ruined fortresses, indisputable traces of what used to be irrigation systems, and some surviving artifacts. History gives witness to the fact that several hundred years ago these regions had flourishing towns and villages full of gardens and green fields, and that life there was quite active. Archaeologists maintain that those inhabited territories were abandoned because of wars and the destruction of irrigation

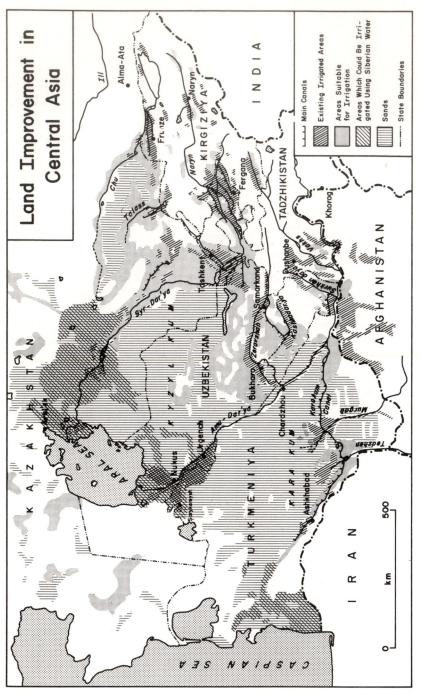

Figure 9.1. Land improvement in central Asia. From Reteyum, 1976.

structures. With the abandonment of the inhabited districts, the land was left open to deflation processes (wind erosion) which developed rapidly and extensively. For a long time, the development of wind-related processes was thought to have been attributed to the "progressive drying of central Asia." However, scientific investigations have shown that desolation of extensive territories stemmed not so much from weather changes but from social processes. The reverse, however, is also true. During the years of Soviet power, many areas that were "dead" or unproductive deserts only a few decades ago have been turned into flourishing districts of developed industry and agriculture.

The Communist Party and the Soviet government have shown a consistent interest in the reclamation and transformation of deserts, an interest connected with the vital needs of economic and social development of the U.S.S.R. in general and of the republics of central Asia and Kazakhstan in particular. Urgent measures were taken to train skilled specialists while well-equipped expeditions guided by eminent scientists were sent to study the deserts.

Gradually, desert territories were drawn into use by industry and agriculture; towns and villages were built there, as were canals and roads. All these large-scale economic measures, followed by fundamental investigations, caused the U.S.S.R. to change radically its previous position concerning the alleged limited natural resources of the deserts of central Asia and Kazakhstan.

Reclamation and efficient utilization of the deserts' natural resources were employed to deal with such problems as water supply, road and settlement construction, and the fixation and afforestation of sandy regions.

Soviet scientists studied natural conditions of the deserts and carried out many experiments and commercial projects which allowed them to discover and to develop a number of promising concepts of theoretical as well as of economic importance.

At present, the U.S.S.R.'s deserts have received more study than most other similar regions in the world. To solve the problem of efficient utilization of natural resources and to decide on the location of economic centers in any geographical zone, one must first carry out comprehensive investigations necessary for the classification and the estimation of natural conditions and resources. Our country has worked out methods for such investigations, methods which result in a rather quick assessment of all natural processes and resources and in finding the best approaches to the reclamation, efficient utilization, reproduction, and conservation of those resources. At present we have a unified classification of desert types which takes into consideration diverse conditions as well as the possibilities for the economic reclamation of desert areas. This classification was obtained by applying different methods based either on one principal natural (and/or economic) feature or on several interconnected features.

In recent years the natural conditions and resources of desert lands have been studied not only by on-ground methods but also by remote sensing on a large scale, a method that contributes to quality, while reducing time and expenses. This has led to a theoretical background that is now being worked out and applied to a complex interpretation of aerial photos. Such techniques for special mapmaking are based on aerial surveys and are currently used on a commercial scale. A study has been made, for example, of the spectral reflectance of some desert landscape features and of suitable seasonal periods for aerial photography. Photo-interpretation keys of the major desert types have been compiled. In addition, a new landscape method of aircraft and space photo-interpretation is in use. This method is based on an integral estimation of natural conditions, and flexible interpretation and extrapolation of various data.

Water Supply

Water supply is still the primary problem in desert regions. Millions of hectares of new land cannot be drawn into active industrial and agricultural reclamation schemes unless this problem is solved.

The basic water supply in the deserts comes from local waters such as fresh underground water and runoff water. As a rule, such waters are used at the place of their origin, but progress in equipment development and in methods for drilling wells and pumping water, together with the availability of a more reliable power supply, have made it possible to deliver water over considerable distances. Large water supply lines, for example, in southwest Turkmenistan and in west and central Kazakhstan now bring local water to towns, mining areas, and industrial centers.

Unlike river water, local waters are dispersed over vast territories, are located in specific places, and have a relatively limited discharge which usually is adequate only for the small farms and enterprises scattered across the desert. In many areas the total amount of local water is enough to supply important consumers with water, provided that such water is efficiently collected and stored. Scientists, for example, have calculated that during a year of average annual precipitation, the deserts in central Asia and south Kazakhstan receive four times as much water in the form of rainfall as flows in the Amudary River (about 60 km³). However, potential sources of local water continue to be used inadequately, despite the fact that there has been widespread use of fresh underground water which has been discovered over considerable territories in the deserts in recent years.

There are many ways to harvest rainwater in the desert, the most efficient being to store runoff water in natural underground strata, where fresh rainwater drains down to form a kind of lens floating on top of the subterranean salt water (Kovda, 1961, p. 195). This problem has been thoroughly studied for the Karakum area, where zones of aeration occur in sand beddings within the water-containing rock strata. Such underground storage lakes guarantee an economically reasonable year-round supply of fresh water for cattle in the most remote pasture lands.

However, takyr (clay) storage lakes do not exist everywhere. Thus, in such regions it is possible to use small water storage basins made of asphalt and cement. At present, the use of new lightweight, heat resistant, antifilter materials are being considered to form a strong water-tight surface cover for such basins. With such materials—either synthetic films or liquid emulsions— one can construct water storage basins in any region and in this way collect a quantity of rainwater and plunge it into the zone of aeration, thus making an underground storage lake.

The natural annual runoff in takyr soils averages about 300 m³ per ha, compared to 700 to 800 m³ under a controlled runoff-retention system. There would be enough water to supply a flock of 800 sheep year-round with fresh water mixed with mineral water.

With respect to mineral water, the U.S.S.R. has recently achieved some degree of success in using such mineral underground drainage as well as sea waters for irrigation. For example, irrigation with mineral waters, with a salinity of up to 5 to 6 g per liter, allows a variety of crops such as sorghum, maize, sunflowers, sudan grass, and rice to be cultivated with good yields on a reasonable economic basis.

Given that mineral water resources in Soviet deserts are practically inexhaustible and that with reliable technical equipment one can make good use of them, desalination of such waters will be very important in improving the water supply of small consumers dispersed throughout the deserts. In fact, much progress has been achieved in the development and construction of large stills for industrial and communal water supplies in the U.S.S.R. In particular, in Shevchenko (Kazakh SSR) and in Krasnovodsk (Turkmen SSR) large stills have been constructed to supply important industrial centers.

Although small stills, especially mobile ones, have not been developed extensively, the home industry turns out small stills with a capacity of 5 to 25 m³ per 24 hours for conversion of saline water with a salt content of 5 to 8 g per liter. But these installations accumulate scale and are either inapplicable or of little use in deserts. Therefore, in certain regions of the desert where there is no fresh underground water or where there are no water-storage basins, people are supplied with water by motor vehicles, planes, and helicopters. This increases the cost of water considerably and makes it impossible to develop pastures for which one needs cheap water to keep stock breeding profitable.

Finally, water desalination by freezing draws some interest, too. Although the effectiveness of this approach has been proven experimentally, it can only be used in the northern regions of the desert zone, where air temperatures drop below zero and thus favor the formation and the long-term storage of ice.

Irrigation

Climate and soil conditions in irrigated desert regions in the U.S.S.R. allow for the cultivation and good harvests of various crops. Extremely arid (desert) conditions combined with the large quantity of solar energy, can provide very stable agricultural conditions for cultivating highly valuable crops. In fact, irrigation farming in the desert zone is the most profitable kind of farming. Economists have estimated that the per-hectare yield of irrigated land in the central Asia Republics can be ten times as much as that under nonirrigated conditions.

The present area of irrigated lands in central Asia and southern Kazakhstan comprises about 6 million ha, while the area of land where irrigation is feasible is a dozen times greater. But even if the irrigation development plan is implemented in the coming years, there is yet another vast area of potentially irrigable desert land of not less than 200 million ha that present and future generations should use to their best advantage.

At present, a vast quantity of data has been gathered concerning the question of whether other Soviet desert areas would be suitable for irrigation farming, and concerning ways to improve land reclamation of older irrigated areas. For example, research data on the water-salt composition of soil has made it possible to forecast the influence of irrigation on the processes leading to secondary, or increased, salinization of soils. Such data have also revealed the extent of the impact of century-old irrigation on the process of soil formation.

Investigations over several years have revealed the genetic and paleogeographic peculiarities of desert soils and have determined their similarities and dissimilarities to the soils of neighboring deserts, thus making possible the formulation of a unified classification of desert soils of the U.S.S.R. As a result of these and other investigations, scientists have proven that a differential approach is required in order to reclaim virgin desert lands. Different methods of soil reclamation must be applied to suit the variety of natural and social conditions—the latter based on historically determined economic specialization.

Usually, lands of plain relief, which require little planning, are suitable for irrigation. This was made a strict prerequisite in carrying out intensive research in agrosoil science and hydrogeologic amelioration for designing man-made irrigation schemes. Lands of this kind near water sources, however, have already been reclaimed. Large canals have been built to deliver water over great distances at considerable cost with the help of modern equipment. Meanwhile, there are many other near-oasis sandy areas where simple planning and proper watering can provide for the development of oasis agriculture. Such sands in the U.S.S.R. desert zone occupy an area of about 2 million ha. The reclamation of these sands is rather inexpensive, since they are close to already reclaimed lands as well as to existing irrigation systems. In fact, the reclamation of such sands often costs less than delivering water over long distances.

The Desert Institute of the Turkmen Academy of Sciences in Ashkhabad has determined that by applying normal dosages of organic mineral fertilizers to serozem-meadow soils (grayish desert soils) and by watering them with an average of 500 m³ per ha, high yields can be achieved of *sorgho Turkmenskoye* (900 double centners—900,000 kg—per ha), Orange-160 (600 double centners per ha), corn (500 double centners per ha), watermelons (300 to 500 double centners per ha), and pumpkins (200 double centners per ha).

The primary cost for one centner of green matter, for example, fodder crops, is currently half as much as that for the same crop grown on standard agricultural lands. The reclamation of aeolian, or windblown, sands greatly helps to reduce wind erosion because the root systems of the plants, penetrating 15 m deep into the soil, stabilize the otherwise loose sands.

More than 10 million ha in the U.S.S.R. desert zone are takyr lands that are considered barren and useless for agricultural and for reclamation purposes. However, work on takyr soil reclamation using rainwater has been carried out in Turkmenistan, where experiments have shown, for example, that high-stem stands, fruit trees, wine grapes, and melons can be grown on these soils with the construction of two trenches: a water retention trench and a plantation trench, with depths of 30 to 35 cm. However, the coefficient of land usage in such areas is relatively low (15 to 20%) because the land between the trenches is reserved for rainfall harvesting and must remain unused. These problems need further study. However, it is already clear that with the help of such mechanisms, even without artificial watering, large areas of takyr tracts can be converted into agricultural lands that will yield a high economic profit.

Pastures

The main requirement of Soviet karakul (sheep) and camel husbandry is the availability of large natural desert pastures. Given that the primary cost of desert animal produce—meat, wool, and karakul—is 50% lower than the average of those products in other parts of the U.S.S.R., and that the total area of the Soviet desert pastures is more than 170 million ha, there is great potential for raising animals in these desert regions. Compared to some African and central Asian deserts, Soviet lands have stable natural fodder which, because of their more regular precipitation, can be used year round.

Plant Cover

The plant cover of deserts, the genetics of plant cover formation and development, and the similarities and dissimilarities of desert plant cover in different countries have been well studied. Pasture maps of different scales have been compiled and yearly variations of fodder yields in different types of deserts have been estimated.

Research concerning the structure of natural and artifical phytocoenoses (plant communities), as well as the biological and economic aspects of plant productivity, are of particular interest. For example, important studies concerning the improvement of natural pastures by means of plowing and sowing native feed plants—species of *Artemisia, Kochia, Haloxylon, Salsola, Calligonum,* and others—are being undertaken. Further, it has been found that the productivity of natural pastures increases 20% after afforestation of sands, and that these pastures can be used for 25 to 30 years with little or no additional care. Finally, if scientists attempting to forecast fodder stocks are successful, they will be able to plan in advance the efficient use of natural desert pastures.

Blowing Sand

The problem of combatting sand drift and deflation is one of the most important obstacles to the economic development of sandy deserts. As is now known, the local populations in earlier times suffered greatly because of migratory sands. Thus, the nature of deflation processes must be kept in mind when carrying out construction projects in deserts.

Soviet scientists have studied the basic characteristics and effects of aeolian action and have developed useful measures for controlling it by using such techniques as matting, astringent substances (oil wastes and polymers), nonaccumulative sand movement, afforestation, and the like.

Solar/Wind Energy

Solar and wind energy resources are practically inexhaustible in the desert. That is why electrification and other uses of these resources can play a major role in desert land reclamation. For example, scientists show that one can convert up to 7% of solar radiant energy into thermal energy and up to 10% into electrical energy with the aid of photo- and thermo-elements. In 1975 a semicommercial, hothouse-like, solar distilling installation was built on a state-owned karakul sheep farm in the Turkmen SSR. The capacity of the still is 1,800 m^3 per year of fresh distilled water. With proper care and maintenance it can be used both as a saline water still and as a rainfall-retention device.

Power for the pumps that deliver water from the wells is supplied through a solar-powered plant with a capacity of 0.5 KW. The generator converts radiant energy directly to electrical energy. Using such a method, the cost of 1 m^3 of fresh water is 2 roubles, 10 kopecks ($2.79 at 1976 prices). Scientists believe that the construction of such stills in desert regions is economically reasonable.

The potential application of solar and wind energy to isolated, small desert farms and settlements is very promising. There are many purposes for which solar energy could be widely used, such as: cold and steam production, water boiling, water distillation, fruit drying, and bleaching. Some solar and wind-powered installations have already been tried as water heaters, solar stills, solar kitchens, wind-powered water-raising devices, electrical generators, and others. At the low efficiencies required by small plants, their source of energy is great and reproducible. Solar energy can also be used to raise the productivity of crops. Experiments have shown, for example, that solar greenhouses can be successfully used to produce valuable agricultural crops year round without requiring any additional heating.

Nowadays, raising water from deep wells is accomplished by pumps. However, the introduction of mechanized water-raising devices in remote areas of desert regions is rather expensive in terms of manpower requirements and of fuel for diesel engines and electromotors. Significant savings can be gained by using wind as a source of energy. Research-oriented experiments have proved that wind-powered motors are reliable sources of energy and can replace the laborious work of men and animals (camels), provided the motors are properly maintained.

Finally, of special importance is the task of building steam-cycle greenhouses that are nearly closed systems in which evaporated and transpired moisture can be accumulated on the inside roof and in ancillary condensers to be used again for watering.

Conclusion

Scientific and technological advances have made it possible to develop mineral wealth on a wide scale and to improve irrigated agriculture and range stock-raising practices in the desert. During the past 10 to 15 years, reclamation of desert lands in the U.S.S.R. has been proceeding at a very fast pace with the construction of large, navigable irrigation canals, storage lakes, and dams. In the past 10 years, for example, the irrigated areas of central Asia have increased by half a million ha because of the radical reconstruction of the irrigation system and the construction of large canals, such as the Karakum, the Karshi, and others. As a result of such constructions, numerous extraction and processing enterprises, modern towns and highways have been built throughout the desert. Nevertheless, the water supply in arid and semi-arid regions remains a crucial problem, and the hydro-resources of the Siberian rivers are under consideration with a view toward transporting water to southern areas of the U.S.S.R. Implementation of such a monumental undertaking is quite possible considering Soviet experience in irrigation engineering. Water transportation from Siberian rivers to Kazakhstan and central Asia will resolve one of the immediate, cardinal problems of desert land development: water supply.

Experience has shown that the reclamation of deserts should be approached with a full knowledge of the nature of the area under investigation; otherwise the probability is high for undesirable consequences that are more likely in desert regions than anywhere else. Deserts are particularly delicate, changeable, and extremely susceptible to man's activities. Their nature has been influenced more than other relief forms by anthropogenic factors. Their vegetation is poor and their soils readily deflated. In addition, woody shrub plants are easily damaged and grow back slowly. In nondesert territories tracked and wheeled vehicles cause little damage to soils. In deserts, however, such vehicles tend to cause deflation by damaging sand-binding plants and soil cover. Desert ranges—which only very slowly restore their grazing capacity—are easily destroyed by cattle through overgrazing and trampling. In this connection, it is important to remember that the vegetative cover of deserts must be seen as a means for stabilizing drifting sand. Careless destruction of desert vegetation will create damage

10 times greater than any profit that might be realized, for example, from the exploitation of its sources of wood.

It is important to note that any anthropogenic intrusion into nature will inevitably change the environment, sometimes radically, as can be seen at first sight in the vicinity of oases and industrial settlements. The incredible extent to which man, with the aid of modern technological methods, has debased and continues to debase the environment has become increasingly apparent.

The concepts of a reasonable "give-and-take" development and of a conservation of desert resources, while still disputed, are clearly indispensable prerequisites for the proper functioning of desert ecosystems. Rational policies of desert land reclamation should be based on a proper understanding of structural regularities of primary and secondary productivity and of the biological rotation of substances in geobioses (ecosystems).

In the U.S.S.R., the desert and soil resources of central Asia and Kazakhstan—resources which constitute the national wealth and property of these republics—have still been neither adequately described or estimated. Such lack of information often results in mismanagement and misuse of these lands. Excessive cuttings of shrub vegetation and the uncontrolled grazing of livestock are still evident, some plots being overused and others underused. Inadequate attention has been given to range improvement. A rational desert grazing economy, for example, calls for a total mobilization of all the available resources: complex application of grazing and stored fodder, laying in of natural and artificial fodder, introduction of a rational grazing system, and an intensification of range sheep production based on enclosed sown pastures. With respect to agriculture, swamping and salinization of soils occur here and there as a result of anarchic land use.

Scientifically-based investigations of ecosystems and the efficient use of desert land within the national economy are still of major concern. The existing stations and their personnel are the solution to some important problems, but not for the entire geobios. In this age of scientific and technological progress, man approaches a new step toward freedom in his interrelations with nature by harnessing its elemental forces while at the same time developing an attitude of respect for them. Scientists must now concentrate on determining the optimum limits of harmonic development of society and nature. We know of many examples in which rapacious exploitation of natural resources by one generation had to be paid for by another.

Scientific and technological progress, combined with the advantages of a socialist system, facilitate the large-scale industrial and agricultural development of deserts and the inclusion of valuable desert resources into the national economy on an industrial basis.

One should not forget that the future availability of natural resources will become progressively more dependent upon man's ability to improve his land-use policy. In this respect, long-range forecasts and the control of natural processes become major concerns in the rational management of our natural resources.

Appendix: Abstracts from *Problems of Desert Development*, 1976, volumes 3-4 (Ashkhabad, U.S.S.R.)

M. P. Petrov, *Obstacles to the Development of Desert and Semidesert Areas and the Conservation of Their Nature*

There is an increasing evidence that human society, if it continues as at present, and the rapid advances of technology will produce "ecological sicknesses" in arid lands, each adding to the development of desertification and affecting dangerously the biological productivity and economic activities of man. The two courses through which desertification is developed are: natural progressive drying of desert territories and their irrational exploitation by man. Coincident effect of the aforementioned factors could lead to catastrophic results (the drought in Sahel, 1971-73). Desertification in semiarid regions hastens man-made desert occurrence, and in arid regions—extra-arid desert progression. The reasons, rates, and scales of desertification are different in every country. As the relationships between certain landscape components in arid regions are unstable, their reproduction occurs rather slowly. Desertification begins the moment when the degree and rate of anthropogenic influence on arid lands exceeds the reproducing ability of landscapes.

To avoid the unfavorable effects of man on the unstable arid ecosystems it is necessary to elaborate the effective measures of planned and scientific approach towards rational and proper technological exploitation of arid land natural resources. Desertification in the U.S.S.R. advances at low rates and rarely reaches a critical point. We are only now seeing the deterioration of pasture lands as a result of overgrazing or development of sand drifts in near-oasis regions and settlements. Scientific long-term prognostication of landscape dynamics based on comprehensive physical and geographical surveys is fast becoming an indispensable tool in elaborating effective measures aimed at desertification control and rational use of desert natural reserves.

A. G. Babayev, *Problems of Desert Land Comprehensive Reclamation and Their Solution in the U.S.S.R.*

The successes of science and high level of technology inspire the expanded development of yet extensively exploited and poor habitable desert lands.

The total area of deserts in the U.S.S.R. is 300 million ha. The deserts, for the most part, are located in the territories of Kazakhstan, Uzbekistan, and Turkmenistan.

The hitherto methods of desert land reclamation were primitive, the rates of the development—slow. The past 20-25 years have seen a noticeable improvement in the exploitation and reclamation of desert territories, which are widely involved in the sphere of industrial and agricultural use. Striking illustrations are the building of great irrigational and navigable canals, water-stock basins and dams, vast areas of desert lands involved in humid farming, towns and settlements built in deserts, new trade and stock-raising centers, highways and power transmission lines. Desert became the source of obtaining diversified agricultural products and industrial raw materials. The present achievements in this sphere and large-scale use of scientific developments lend a new acuteness to the problem of further amelioration and rational use of desert natural resources.

Among the chief leading trends of scientific developments which have direct effect on desert exploration are—complex physicogeographical investigations by virtue of different methods and approaches; problems of water supply and reclamation; land resources, their rational use; intensification of animal farming and strengthening of forage base; forestry and phytomelioration; wind erosion control; application of solar and wind power; mechanisation of laborious processes in desert amelioration; adaptation of man and animals to climatic conditions of deserts; substantiation of economic efficiency of different methods aimed at rational use of desert natural reserves.

Much work has been done by the Soviet scientists to study the nature of desert. It allowed for the scientific forecasting of the promising scientific and technological trends having direct effect on practice.

Yet, there is no commonality in approach, and therefore in solution, of the fundamental issues. Some of them have already been solved and are used in practice, others are under elaboration, still others remain open and much-debated problems.

N. T. Nechayeva, *Influence of Anthropogenic Factors on the Ecosystems of Middle Asia Deserts*

The incredible extent to which man has debased the environment causes rapid and irreversible changes of desert ecosystems. Proper estimation of the effect of grazing is dependent on man's being conversant with biological features of plant life forms, the structure of phytomass of plant associations, and other factors. Vegetational cover of Middle Asia includes various life forms of plants: trees—1.8%; shrubs—11.5%; semi-shrubs—14.1%; perennial grasses—21.6%; biennial plants—0.8%; and annuals—50.2%.

Desert ranges of Middle Asia are subdivided into shrub, dwarf undershrub, and herbaceous ones. Shrub ranges are considered to possess the highest grazing efficiency and diversity of fodders. The herbage yield of all the types of ranges averages to 4.5-5.1 c per ha, but its annual variations are different.

The herbage yield of shrub ranges, in particular, varies between 3 and 7.5 and of herbaceous ranges—between 1.3 to 12.3 annually.

The effect of grazing upon the vegetational cover of ranges is another major area of study concerned with the influence on desert environment of anthropogenic factors. Under the mean rate of stocking, the general palatability of the herbage will be 66 to 51% of total phytomass of ranges.

Overgrazing with systematic removal of more than 75% of annual sprouts hastens deterioration of ranges, while the prolonged absence of grazing was established to exert an unfavorable influence on desert vegetation. The 4 to 5 year resting of range lands has been established to decrease the forage yield by 20%. Repeated grazing of ranges in spring hastens (in 3 to 4 years) the depression of Carex physodes and spring annuals.

Systematic summer grazing of ranges results in excessive loosening of the surface sands.

Winter grazing decreases the yield of annuals. The effect of the autumnal grazing is of intermediate nature between the summer and winter ones, depending on the amount of precipitation.

With the regularities of grazing effect upon the herbage yield in ranges known, practical recommendations can be set forth to develop rational schemes for the rotation of range land grazing.

Proper utilization of natural biological resources will be a more important contribution than ever in the prevention of desertification in arid regions.

A. K. Rustmov, *Wildlife of Middle Asia and Kazakhstan Deserts, Its Utilization and Protection*

Arid lands of Middle Asia and Kazakhstan have all the natural resources for obtaining a great variety of agricultural products, such as karakul, wool, furs, meat, fish, medical raw material, fuel, etc. The resources, tremendous though they are, will be greater if mineral reserves of arid lands (oil, gas, water, and others) and solar energy as well, are taken into account. In this respect the potential of arid lands is destined to play a particularly important role in the economics of the country.

In order to use the resources of arid lands to their full extent today and in the future, controlled exploitation of arid lands is being carried out. Comprehensive development of desert regions should not ignore the problems of nature conservation. To this end a number of nature conservation decrees have been accepted in Kazakhstan, Uzbekistan, Kirghizia, Tadzhikistan, and Turkmenistan.

The successful utilization of desert resources will be impossible if the problem of irrigation and water supply is not solved. It is irrigation that will facilitate the transformation of low productive ecological systems into more productive ones. At present, vast programs of irrigation construction are being successfully carried out in arid lands. In this connection, the annual harvest of raw cotton increased up to 8 million tn.

No less important is the comprehensive introduction of chemical fertilizers. But their improper application decreases the productivity of controlled ecological systems, impairing desert wildlife. Thus, elaboration of biologic methods of agricultural pest control becomes necessary. Man's debasement of the environment has dictated an unparalleled need for rational exploitation of natural resources in order to minimize the anthropogenic negative effects.

Irrigation of arid lands exerts a favorable influence upon their nature: the occurrence of new areas of hibernation of birds and animals, improvement of forage resources and watering places. Controlled technological and agricultural exploitation of arid lands will be a major factor contributing to the conservation of desert nature.

B. V. Vinogradov, *Desertification Forms as Shown on Aerial and Space Photos*

Desertification comprises both long-term successions of landscapes in which subarid semidesert and savannas alternate with deserts, and short-term exogenous successions. According to the genesis, desertification can be climato-, hydro-, geo-, morpho-, and anthropogenic. The majority of the aforementioned types of desertification can be identified on the aerial and space photos. The most reliable identification criterion of desertification is a comparative analysis of the data obtained during the repeated photography of the same territory. The latter allows for identification of vegetationless sands, damages of vegetational cover, drying of solonchak soils, etc.

The spectral reflectance coefficient is considered to be one of the major distance keys of desertification phenomena. Its value is inversely proportional to the projected vegetational cover at the decreased rate of soil moisture and increased concentrations of dust in atmosphere. Local forms of desertification can be determined on the aerial and space photos with 100 to 300 m clearance. Spots and strips of different tonalities are the principal images of the aerial forms of desertification on space photos. For instance, strip and zonal images of desertification can be identified on space photos of preoasis landscapes along the rivers of Amudar'ya, Tiger, and the Khiva oasis. Spot images are common for flat range territories of Egypt, Afghanistan, Mali, and Syria.

N. G. Kharin, *Use of Remote Sensing to Study Desertification Processes in Arid Lands*

Arid lands of the U.S.S.R. possess great natural resources. Every year new areas are reclaimed, pipelines are constructed in the desert, new industrial centers appear. Unfortunately, information on the change of natural conditions of the desert is often delayed, and sometimes scientists and managers do not know the current status of the desert.

In our age of technology the problem of studying the natural resources of desert lands can be solved on the basis of remote sensing technique. Taking into account the phenological features of desert plants, the author has determined the optimal seasonal time of aerial photography in sandy deserts of the U.S.S.R. In sandy desert, spring is the most suitable season for aerial photography. Ten days after Haloxylon persicum begins to grow is the optimal time for aerial photography. The end of the period is indicated by the withering of Carex physodes, a widely distributed ephemeroid in sandy deserts.

Panchromatic film is the most suitable film for aerial photography in deserts. In the contact zones between the desert and oases, in river valleys and reclaimed agricultural areas, infrared and false color films give good results.

Photographic films have a comparatively wide zone of spectral sensitivity. They cannot register contrasts between natural objects in narrow spectral bands, but there is another method which provides the image in narrow spectral bands: multispectral aerial photography. The most suitable spectral bands for multispectral photography are: for sandy desert—400 to 525 nm and 600 to 674 nm; for cultural landscapes—600 to 675 nm and 750 to 850 nm.

On the basis of remote sensing special maps are compiled. For nature conservation the following types of maps should be mentioned here: wind erosion maps, maps of sand types suitable for afforestation, maps of forests damaged by insects, maps showing the distribution of harmful substances and salts in water and soil, etc. These maps can be used to plan the protective measures for the conservation of natural resources.

L. E. Rodin, *Primary Productivity of Desert Communities in North Africa and Asia*

Euxerophytic "desert trees," shrubs, dwarf undershrubs, and sod soil grasses are the principal edificators of desert plant communities. They produce the bulk of organic green matter. Besides, cohabitation of the aforementioned plants with ephemeroids and ephemeres has an increasing effect on the organic matter production, especially in years of abundant precipitation.

To analyze the parameters of productivity, data have been taken from 20 geographical points located within 2-103° W.L. and 30-40° N.L. A total of 50 plant communities characteristic and widespread in the regions were considered under investigation.

The present aim for the improvement and conservation of highly productive, viable, and naturally regenerated biogeocoenoses of arid zone is the development of plant communities with tree-shrub biomorphs with deep-penetrating and powerful root systems, and populations of other biomorphs (dwarf undershrubs, ephemeroids, annuals, and perennials) with different periods of growth allowing for the multi-layer (over- and underground) structure.

Z. G. Freikin, *Economical and Geographical Problems of Desert Land Reclamation*

The estimated total area of arid lands in the U.S.S.R. is 14% of the total territory. The influence of arid lands on the economics of the country is large, as are the difficulties of using the natural resources of arid land.

Historically, desert territories have been economically developed in three ways: (a) animal husbandry—seasonal stock grazing—the oldest; (b) agriculture—transforming the desert by irrigation; and (c) industry—extraction of minerals.

The natural conditions and the resources of deserts exist in different quantitative and qualitative proportions. Thus the economic effect of their development is greatly dependent on the way they are reclaimed. It is advantageous to separate desert land into regions according to the rate they are developed, using cost as a measure.

In this way the most efficient procedures for amelioration of the different territories can be determined. In addition, the output of underdeveloped arid lands can thus be raised with the least outlay.

A. P. Lavrov and N. S. Orlovsky, *Soil and Climatic Regionalization of Turkmenistan Plain Part*

Natural conditions are a major factor in determining the management problems of farming. Planning and improving agricultural efficiency requires careful study of natural resources and soil and climatic conditions.

More than 80% of the republic's territory are plains with specific soil and climatic conditions. Regions are distinguished by the following taxonomic units: province, region, and district. The district is distinguished on the basis of a complex of related natural factors: soil, relief forms, lithology, ground waters, etc.

The whole territory of the republic is referred to as the Touran desert province which, in turn, includes 12 soil-climatic regions. They are subdivided into 58 districts, each characterized by comparatively similar soils in view of their origin and agronomic indices. The latter presupposes the use of effective soil reclamation.

In addition, vast tracts of takyr-like and light serozem soils are scheduled to be reclaimed in zone of the Karakum Canal and bottomlands of the Amudar'ya river.

M. R. Nikitin and M. W. Sanin, *Mineralized Ground Water as a Source of Water Supply in a Desert*

In the deserts of Soviet central Asia, Karakum, and Kyzylkum, where the lack of fresh water is acute, there are large artesian basins, the Amudar'ya and Syrdar'ya, which contain great resources of brackish and saline ground water. This water can be used as it is or after desalting for the rural water supply. The best developed methods for desalting ground water with a higher dissolved solids content are electrodialysis, reverse osmosis, and solar distillation. An analysis of the data on separate desalting plants shows that the cost of water obtained by electrodialysis within a capacity range of up to 1,000 m³ a day and a content of dissolved solids of up to 1 gram per liter is 100 to 30 kopecks per m³ and specific capital investments are 500 to 200 roubles per m³ a day. As for solar distillation, with a plant capacity of 2 to 12 m³ a day, the water cost amounts to 400 to 200 kopecks per m³ and specific capital investments approach 6,000 roubles per m³ a day. Studies have revealed zones in desert and semidesert areas where it is advisable to use only desalted ground water. With a desalting plant capacity of 1,000 m³ a day, the maximum distance of profitable supply of fresh water—which may be the criterion for creating such zones—is 11 to 17 km, depending on the method of water supply and treatment.

V. N. Nikolayev and A. A. Amengel'diyev, *Principles of Desert Land Classification According to Quality*

Land classification according to quality, one of the chief elements of soil cadastre, utilizes the scientific assessment of soil natural fertility. Fertility and nutritive value are primary concerns in the classification of grazing land.

Taxation of grazing lands differs from taxation of arable lands and requires special methods. Forage productivity—expressed in conventional forage units measured by digestible protein—is considered the basis of grazing land classification. The highest point of a regional 100-point classification scale for desert zones should be equal to 1,000 conventional forage units (c.f.u.). Thus, each point will be 10 c.f.u. The regional classification scale should be in accordance with the All-Union one, allowing for the development of a coordinated soil cadastre. For the qualitative evaluation of grazing lands in the U.S.S.R., including controlled grazing and haylands, the top point of the scale should be increased by ten-fold.

The territory of Western Turkmenistan (above 10 million ha) was used as an example of the classification of desert rangeland. Mean annual values have been used for the taxation of the lands under investigation. The major portion (8.7 million ha) of the investigated lands is represented by depleted ranges approaching 4.0 to 13.8 points. 1.5 million ha of sparsed mountainous tracts are evaluated up to 38 points. Only 60,000 ha of river valley pastures are assessed on the average of 64 points and are called the mean category of range lands.

E. A. Vostokova, *Landscape Map of North Africa and Its Hydrologic Interpretation*

Since the problems of water supply in the arid zone of North Africa are of prime importance, a landscape map of North Africa has been compiled to estimate the possibilities of landscape-hydrologic analysis for ground water prognostication.

The hydroindicatory analysis was first applied with a basis of small-scale mapping of vast areas. The practical purpose of the landscape map influenced the legend construction. For instance, detail mapping has been made of "hydrogenous" natural and territorial complexes with ground waters of shallow occurrence. For mapping land-scale types, different-scaled topographic and thematic maps have been used. To compile a landscape-type map a multistage legend was elaborated on the basis of the analysis of physicogeographical, tectonic, morphostructural, and climatic features.

The compilation of this type of map for North Africa does not take into consideration the anthropogenic influence of landscape breaks.

For the subsequent hydroindicatory interpretation of the map, landscape-hydrologic regionalization of North Africa was applied with regard for the data on depths of ground water occurrence, the rate of mineralization, and the location of wells.

Despite its speculative nature, the map presents general information about the first aquifer of the territory under investigation.

M. P. Aranbaev, *Ancient Oasis Soil Ecosystems in Connection with Desertification*

Ancient irrigation soils in oases formed on a thick series of the agroirrigational strata are among the components of soil macrostructure in desert zones. The genesis of ancient oasis soils is related to the occurrence and development of ancient agriculture, and the anthropogenic effect on natural ecosystems in desert zones is characteristic of the aerial of ancient oasis soil formation.

The primary productivity of the ecosystems and the structure of the organic mass under the ancient oasis landscapes changes considerably due to the different structure of soil cover of the agroirrigational relief forms and the difference in potential productivity of the soils of certain units of the agroirrigational complex. High potential productivity is caused by the slight weathering of alumosilicate, the high specific weight of thinly dispersed masses, and the thick humus shell. The specific microaggregate structure of the solid phase of ancient oasis soils promotes the optimal hydrophysical properties of the soil processes.

The differentiation of the soil cover, the productivity of agroecosystems, and the efficient application of mineral fertilizers are dependent on the geochemical differentiation of the solid phase of suspended accretions and hydrosaline masses within the agroirrigational complex.

The space variability of the biological and economic productivity of ancient oasis ecosystems exists during their long-term desertification and the following repeated irrigation.

F. F. Sultanov, *Man and Animal Adaptation Under Conditions of Arid Zone*

The reclamation of arid lands with peculiar natural and climatic features immensely increases the importance of research on the physiological adaptation mechanisms of man and animal organisms. The aim of such research is to look for ways to improve the quality of life in arid zone conditions.

In the process of adaptation to high environmental temperatures, the first reactions of the organism exposed to new environmental conditions are of primary importance. This is because these reactions, while indicating the intensity of high temperature effects on the organism, cause the malfunction of certain systems. The latter makes difficult, or even prohibits, adaptation to high environmental temperatures. Bearing in mind that adaptation to heat is a temporal process, investigations have been initiated of the primary effects of arid zone factors on nonadapted organisms.

To briefly summarize the characteristics of the processes occurring in nonadapted organisms exposed to high ambient temperature it is necessary to single out the imperfect nature of controlling mechanisms of heat distribution in the organism, the overload of the cardiovascular system at a relative circulatory incompetence, and the inadequacy of hormonal-humoral stress reactions.

The basis of adaptation to arid zone conditions (to high temperature in particular) is in the improvement of regulatory mechanisms eliminating the side effects of unspecific reactions. In the course of adaptation, redistribution of functional activity of different systems occurs, decreasing the load on hemodynamics and adapting the tissue metabolism to new environmental conditions.

The concepts here discussed make it possible to approach the solution to man's adaptation to arid zone climatic conditions.

Z. Sh. Shamsutdinov, *Methods Increasing the Productivity of Desert Pastures*

Low productivity of range lands (1.5 to 3 ha³ of dry matter) and significant seasonal and annual variations of forage crops in Middle Asia dictated an unparalleled need for the introduction of effective methods aimed at the radical improvement of lands.

In this respect, more than 250 species of forage plants referred to as wild flora have been tested in desert and semidesert zones. Of the species tested, Haloxylon aphyllum (Minkw.) Iljin, Salsola paletzkiana Litv., Kochia prostrata (L.) Schrad, Salsola orientalis S. G. Gmel, Artemisia diffusa Krasch. proved to be the promising species.

In culture, the aforementioned shrubs and semishrubs develop a powerful and deeply penetrating root system. They thrive perfectly, and quickly develop sufficient seed productivity. As a result, the formation of the principal phytocoenoses occurs during a period of two to four years, characterized by high and stable productivity. The dry matter yield of the artificial pastures established in arid zones reaches 12 to 15 ha³ as opposed to 1.5 to 3 ha³ in natural ranges. Grazing capacity of cultural pastures increases by 3 to 5 times. With due regard for the biology and ecology of shrubs and semishrubs, and based on agrotechnical tests, effective measures of fundamental improvement of range lands were devised: the establishing of H. aphyllum protective belts to control sand drifts and perennial autumn/winter and summer pasture lands.

The measures developed for pasture improvement are a great help in karakul sheep farming. Nowadays the area of perennial pastures in karakul sheep farms of Uzbekistan approaches 200,000 ha.

M. V. Kolodin, *Water Desalination in the Deserts of the U.S.S.R.*

A successful resolution of the water supply problem in arid zones of the U.S.S.R. requires the parallel development of fresh water transportation by channels and pipelines and the technological realization of mineralized water distillation using different types of stills.

Powerful desalination stills (DS) with 13 to 15 thousand m³ a day capacity have been constructed in waterless desert areas to supply domestic and industrial water. The specific performances of the operating plans and those under project are described below.

The continental waters of vast desert regions of the U.S.S.R. are highly mineralized. Special DS have been elaborated for the mineralized water conversion aimed at supplying water to small settlements, country estates, large farms, and stock-drinking places. Their capacity varies between 12 and 350 m³ a day; their specifications are given.

Water is supplied in the desert by means of mobile and transported DS (8 to 48 m³ a day). Some of them are in serial production.

Consideration is being given to the possibility of water conversion with the aid of wind power. The specifications of wind-powered desalination stills (WDS) have been adjusted to domestic and commercial purposes with a daily capacity of 1 and 6 m³, operating according to two diagrams of wind power accumulation. For ranges 30 km from the water source, the erection of rangeland WDS is more advantageous compared to water transportation with the aid of tank lorries.

R. Bairamov and S. Seyitkurbanov, *Principal Trends in Scientific and Technical Research on Developing a Solar Distiller*

Introduction of a solar distiller to provide water to the desert and semidesert grazing lands of central Asia and Kazakhstan is one of the promising trends in desert grazing economy. To this end, scientific institutions in the U.S.S.R. and abroad are working on developing more advantageous and reliable distillers.

A great deal of careful work has been done to improve the existing projects and to develop new diagrams for solar-powered desalination stills, particularly a solar still section with filmy basin and ferro-concrete-tray basins.

Special investigations have been aimed at the reliable improvement of the still units. Also undertaken were the tests of combined application of solar energy, natural freeze, geothermal water, and waste heat of power installations for water conversion.

The results are given of annual tests of solar still units to be applied in stock-drinking places. The pros and cons are described of the existing diagrams of stills, along with the aggravating effects saline water and distillate exert upon the materials the plants are made of.

V. P. Cherednichenko, *Engineering-Geomorphologic Investigations in Sandy Deserts*

The rapid development of central Asian deserts requires engineering-geomorphologic studies. The specific character of these investigations in sandy deserts is attributed to zonal and regional features of the area. The methods and content of the surveys, and the degree of detailed data processing, are dependent on the type of engineering construction and relief-forming processes. Engineering-geomorphologic investigations deal with the morphology and dynamics of the relief and relief-forming factors and processes. The application of the morphologic and morphodynamic analysis methods is mainly aimed at obtaining morphographic and morphometric characteristics of the aeolian relief and examining the interrelations between the relief and aeolian relief-forming processes. The data on the problem under investigation are a great help in developing nature-protection measures, and can be successfully obtained with the aid of stationary surveys. The engineering geomorphologic map facilitates the technologic evaluation of the recorded phenomena and processes. Based on the geomorphologic data and their engineering evaluation, practical recommendations are developed to determine optimal ways of locating industrial objects in the system of aeolian relief, with due regard for the relief-forming processes and technology of engineering and sand-stabilization measures.

V. S. Zaletayev, *Organism Dispersion and Zonal Types of Desert Biogeocenoses in Central Asia and Kazakhstan*

Biogeocenosis structure reflects the natural zonal regularity; the specific zonal types of biogenic environment are the result of this reflection. The main factor which determines the biogeocenotic structural pattern is the character of the organisms. The phenomenon of biogenic amelioration of the environment is the base for a geographical/cenotic paradox: while the north-to-south aridity of climate increases, the extremity of the environmental life decreases. The pattern of organism dispersion acquires a qualitatively new form; the dispersion of individual organisms is replaced by that of ecologically associated groups.

The author calls the northern desert type of organism dispersion "individualizing dispersion" and the southern desert type "aggregating dispersion," that is, the dispersion of small, isolated, but functionally associated organism groups.

The typical zonal form of southern desert cenotic structures may be called the system of input structures.

The essence of it is in the development of microgroups round the cenotic formation of organisms. Every group can possess its own morphological structure.

The dispersion types and biogeocenotic structure patterns are the important zonal environmental characteristics of arid zones.

Sh. E. Ergeshev, S. U. Umarov, and A. Bazarbayev, *Landscape Peculiarities of Sands in Southern Uzbekistan and Prospects for Their Development*

On the basis of integrated physical and geographical studies, the paper describes the natural conditions prevailing in the sands of southern Uzbekistan and outlines measures for their comprehensive development.

A detailed investigation has been undertaken of the genesis, mechanical properties, and mineral and chemical composition of sand and its nutrient substances. Also studied were wind and water regimes and the dynamics of drifting sands. We have developed a typological classification of sands and outlined their regionalization aimed at their amelioration and agricultural development. As a result, practical recommendations have been set forth for the integrated development of sandy zones. To this end, the diverse and tangible influence of man on sands was analyzed.

The landscape types have been singled out and mapped, including hillock, dune, and ridgy types of sands. They are divided into subtypes: small, mean, large, and high hillock sands. And finally, the paper outlines the scientific and geographical principles of sand reclamation.

L. M. Grave, *Technogenic Material Exchanges and Geocomplexes in the Karakum Canal Zone*

The paper deals with the influence of human activity on the relief forms of the canal zone.

Application of technological methods during construction jobs on the canal (its deepening, widening, draining water excess, etc.) and the interaction between its functioning and desert nature have considerably changed the landscape of the canal zone.

The Karakum geotechnological system is very complicated and heterogeneous. Technogenic material exchange is presented by liquids and solids. Their interaction has led to pulp sands, washouts, etc. The paper gives their classification characterized by types of material exchange.

T. I. Fazilov, *Theoretical Principles on Binder Protection of Railway Tracks Against Sanding*

Over 4,000 km of railway tracks in the U.S.S.R. running across the sandy deserts and semideserts are endangered by drift sands.

For the last 10 to 15 years much attention has been focused on the application of diversified types of binding agents to control drift sands.

The paper cited includes information on a variety of sand binders used in the U.S.S.R. and abroad, as well as data on experimental work accomplished for a period exceeding 10 years by the members of the research laboratory of the Tashkent Institute of Engineers of the Railway Transport.

The control of sanding in accordance with the above-mentioned classification can be accomplished by binders, diluted binders, emulsions, mastics, etc.

The optimum method for control of sand deflation is selected with the help of a special table depending on afforestation in the given district, motion of sand and its properties, deflation processes, weather and soil effects, as well as technical and economic prerequisites.

The criteria for assessment of the service characteristics of mineral sand binders are the weight of the aggregate particles that ensures stability of the layer at the first critical speed and service time of the coating evaluated by frost resistance. The mechanical protection in addition should possess optimum geometrical dimensions.

An appreciable part in the work is devoted to test results (with the help of electron microscope, differential-thermal and roentgenostructural analyses, speeded tests in an artificial weather chamber and aerodynamical tube), as well as to investigations and the introduction into practice of highly efficient binding agents for protection of railway tracks against sanding in the Karakum desert, Caspian lowland, and the Mangyshlak peninsula.

S. L. Gendler, K. Ya. Fedorenko, G. S. Kalyenov, and L. A. Korokhov, *Comparative Analysis of Mineral and Chemical Composition of Sands in Simpson (Australia), Sahara and Karakum Desert Territories*

Mineral and chemical composition of sand obtained from the Simpson, Sahara (North Africa), and Karakum (central Asia) desert territories were investigated. Each one is specific due to the differentiation in the genesis, degree of grading, as well as intensity of chemical weathering of the materials. This differentiation causes great variance in plant habitats through the aforementioned desert territories.

N. G. Kharin and M. P. Petrov, *About Desert Terminology*

Desert terminology includes numerous terms with various origins. In the past, desert terminology included mainly the terms on environmental conditions, but now there are many terms connected with the practical activities of people in deserts. Expressions make their appearance, and different languages contribute to the vocabulary.

The authors of the paper have attempted to create an explanatory desert glossary. The makers of the glossary have aimed at summarizing under one cover the geographical terms of the world's deserts. Included in the glossary are the terms on physical conditions of deserts, local geographical terms, and words dealing with the practical activity of man while reclaiming desert lands (collecting and storing water, grazing, sand dune areas, afforestation, etc.).

"Phytomelioration" is a new term which one can find in the glossary. The term designates the land improvement by the establishment of forest plantations, cultivation of agricultural crops and other plants which play the protective role and (or) give wood, forage, and other products. In the deserts of central Asia the phytomelioration work is carried out at the large extent.

Local geographical terms have been included not only because of their toponimic value but also because of their wide usage in the scientific literature. After being used for the description on some particular local features, these terms may be then widely applied in practice. For instance, such words as "barkhan, solonchak, steppe" and many others have become a part of the scientific terminology in many languages.

The authors hope that this glossary will improve the information exchange among the scientists of different countries.

References

Kovda, Viktor A., 1961. Land Use Development in the Arid Regions of the Russian Plain, the Caucasus, and Central Asia. In *A History of Land Use in Arid Regions*. L. D. Stamp, ed. Paris: UNESCO Press.

Reteyum, A., 1976. Transformation Scene in Central Asia. Trans. D. J. S. Shaw. *Geographical Magazine*, August, pp. 682-86.

10
Interim Report of the South African Drought Investigation Commission, April, 1922

May it please Your Royal Highness:

In terms of a Commission issued on the twenty-fourth day of September and on the second day of November, 1920, by His Excellency the Right Honourable Sir James Rose-Innes, P.C., K.C.M.G., Officer administering the Government of the Union of South Africa, we, Your Commissioners were appointed to investigate and report upon certain matters connected with periodic droughts in the Union of South Africa.

Having reached a stage in our investigations where it is considered advisable to issue an Interim Report, Your Commissioners have the honour respectfully to submit to Your Royal Highness the following Report:—

I. INTRODUCTION

1. The several matters upon which your Commission has been instructed to report, all deal directly with problems arising out of drought losses.

2. While the recommendations contained in this Interim Report furnish, as far as they go, direct replies to the Terms of Reference, it

---

Editor's note: This report was presented to both houses of Parliament "by Command of His Royal Highness the Governor-General." Cape Town: Cape Times Limited, Government Printers, 1922.

has, however, been deemed advisable to refer generally to what may be called the "drought-resisting capacity" of the Union because it is necessary to prepare a foundation on which to construct this and subsequent reports. Moreover, the factors responsible for the serious periodic drought losses in the drier parts would seem to be in operation throughout the Union generally.

3. Your Commission, in conducting its investigations, decided not to study the history of the past very deeply, as that has frequently been done before. Two points, however, seem firmly established: firstly, that a large portion of South Africa was dry long before the white man arrived, as evidenced by the name "Karroo" and by the highly-specialised drought-resisting flora of that region; and secondly, that since the white man has been in South Africa enormous tracts of country have been entirely or partially denuded of their original vegetation with the result that rivers, vleis and water-holes described by old travellers have dried up or disappeared.

4. This drying out of extensive areas of the Union is still proceeding with great rapidity in many portions of the country, and in parts now apparently too dry to permit of their growth, tree stumps of indigenous and exotic trees, would seem to corroborate the human evidence to that effect.

5. It is unnecessary for your Commission to vie with the several writers who have, at various times, with facile pen depicted the gloomy and ghastly future which lies before our country, if we permit these processes to continue. The simple unadorned truth is sufficiently terrifying without the assistance of rhetoric. *The logical outcome of it all is "The Great South African Desert" uninhabitable by Man.*

6. From figures furnished by the Census Department, it has been estimated, on prices obtaining prior to 1914, that the direct losses of farmers during the 1919 drought amounted to not less than £16,000,000.

7. What the indirect losses were it is impossible to say, but they were obviously considerable, and *affected every profession, business and trade*. The losses that have occurred, say, during the last fifty years from these recurring devastating droughts cannot but lead to the conclusion that the position demands the earnest attention, and, if need be, intervention of the State.

8. During the course of its labours, your Commission has found the problem of avoiding losses through droughts one of exceeding complexity, requiring a much more extended investigation than

was at first realised. . . .

9. Much ground has still to be covered, but it is felt that a stage has been reached which permits of this Report being made.

10. The kraaling of stock, occasioned mainly by the jackal, inadequacy of the drinking water facilities, the destruction of vegetation and the resulting soil erosion, which in turn leads to a serious diminishing efficiency of the rainfall, are the main causes of drought losses. That these losses are only a portion of the total losses brought about by these factors, is an additional reason for taking measures against them.

11. At the outset it had to be decided whether your Commission should go to witnesses or witnesses should come to your Commission. The former alternative was chosen as being calculated to give the Commissioners a clearer conception of the conditions and circumstances obtaining in the various parts of the Union; besides which, it would result in the whole problem becoming better known and understood throughout the land. In this decision your Commission had the approval of the Honourable the Minister for Agriculture of that time who, furthermore, suggested that an educative campaign might be simultaneously conducted. Your Commission, therefore, acted in a dual capacity and in its educational work was assisted occasionally by officers of the Department of Agriculture, to whom thanks are due.

12. In notifying time and place of meetings, the Department of Justice (through its Magistrates or Police) and the Department of Posts and Telegraphs (through its Postmasters and Postmistresses), rendered invaluable aid, which is gratefully acknowledged.

13. The course adopted in taking evidence from farmers at public meetings was to request the magistrate, or other prominent citizen, to nominate a number of those present, who would be thoroughly representative of the various parts of the district. These were examined collectively; those not so chosen were given the opportunity to remark upon the replies of the nominated witnesses. On the whole, this plan appears to have worked well; in any case, this method of taking evidence was the only way of dealing with a large number of witnesses in a reasonable time.

14. . . . In all, one hundred and two public meetings have been held, of which Mr. van Reenen attended all; the Chairman, Mr. du Toit, eighty-eight; Mr. Kolbe, sixty-nine; Mr. Gadd, fifty-six; and Mr. Stead, fifty-four. The last two members of your Commission were unfortunately prevented by serious illness from attending a

larger number of meetings. Two meetings duly advertised had to be abandoned owing to swollen rivers, and one owing to an epidemic of influenza. Four meetings were indefinitely adjourned owing to insufficient attendance.

15. Your Commission held its first meeting early in November, 1920, since when, with the exceptions of an enforced break during the General Election Campaign of 1921, and short breaks to allow members to visit their homes, it has travelled continuously up till December, 1921.

16. During December, 1921, through the courtesy of, and the arrangements made by, the officials of the Basutoland Government, your Commission was able to visit the head waters of certain Union Rivers, which have their origin in Basutoland, and study the erosion that is taking place there.

II. RAINFALL

17. Before passing to the consideration of the factors mentioned in the first Chapter as being responsible for much of the devastation due to droughts, it is necessary to state clearly what is meant by a drought in order to prevent misunderstanding.

18. Ordinarily the word means a want of rain, or dry weather, and is perhaps more often applied to crops than to grazing. In this report the term is used with reference to grazing only, particularly for sheep, amongst which class of stock losses have been severest. In some parts of the world a few weeks of dry weather would constitute drought conditions; while over the greater part of the area with which your Commission's enquiries have dealt, as many rainless months in winter would be perfectly normal. Again, within our own borders drought conditions develop much sooner in grass-covered areas than in the scrub-covered Karroo, where the vegetal covering is highly resistant to droughts.

19. The meaning, then, to be attached to the word drought in this report is a period when grazing has become so scarce and the supply of water at the drinking places become so diminished that loss of stock results.

20. From a consideration of the meaning of the word drought we naturally turn to the question of rainfall.

21. Your Commission has not yet been able to complete its investigations in connection with the rainfall factor in the Droughts problem. It is hoped, however, that important recom-

mendations dealing therewith will be submitted when the final report is presented.

22. In the meantime it is necessary to state briefly what the position is. Neglecting the climatic changes of past geological ages, no evidence has been submitted to your Commission to prove that the average annual rainfall in South Africa has changed during recent historic times. Variations from year to year naturally exist. Good and bad years follow on each other with no apparent regularity; but no general upward or downward tendency in the mean annual rainfall can be traced. Unfortunately the accurate gaugings on which such conclusions must finally be based are, for the greater portion of the Union, of very recent date.

23. In studying rainfall and its effects on vegetation, however, we cannot confine ourselves to a discussion of the total annual fall. The useful effects of the rain of any one year depends not only on its amount, but also on other factors. It may have fallen in innumerable small showers and be absolutely useless, or it may have fallen in a few useful soaking rains, or perhaps, again, the entire precipitation may have occurred during a very few violent and devastating thunderstorms. The rain may have fallen early in the summer or perhaps rather more towards autumn, resulting in each case in an entirely different state of affairs as far as the value of the grazing is concerned.

24. This introduces a point on which your Commission has not, up to the present, been able to obtain definite proof. Throughout the country there is a very general tendency among farmers to believe not only that the annual fall has diminished, but also that the rain now usually comes later in the season and that the "good old fashioned" gentle soaking rains are much less common. Your Commission does not consider that definite proof of these alterations has been submitted. The belief is based on personal reminiscences which are particularly treacherous when brought to bear on meteorological data. In cases of this nature witnesses are liable to judge by results alone. The rains of last generation, falling on unbroken, under-stocked grazing lands were more lasting in their beneficial results than rains of equal magnitude falling to-day on veld overstocked, tramped out, semi-waterproof, hard-baked by sun and veld fires. Such rains in the older days were "soaking" rains with long-lasting beneficial results—to-day they are not. The "cloud-burst" is estimated by the layman by the amount of water in the streams. A cloud-burst or heavy thundershower falling on

thickly-grassed virgin veld produced only a trickle in the water courses of former days compared with the torrents rushing down the dongas [gullies] of our eroded sections under similar circumstances to-day. These facts cannot but have influenced the deductions made by witnesses from their personal observations.

25. In view of the foregoing it would not be wise to place too much reliance on evidence of past meteorological conditions where this is based on recollection; and in doing so one does not need to controvert the strong conviction of many witnesses that droughts are now more frequent in occurrence and more devastating in their effect; for it is not so much the actual rainfall as the effect of that rainfall which determines whether or not the season is one of drought. The effect of rainfall is dependent not only on the time and nature of fall; it is profoundly influenced by stock and veld management, temperature and wind conditions. Herein lies the explanation why rainfall records, which take no cognisance of these factors, are frequently at variance with the farmer in the assessment of drought conditions.

26. Whether the character has altered or not, or its quantity diminished, drought losses can be fully explained without presuming a deterioration in the rainfall. Your Commission has had a vast amount of evidence placed before it from which only one conclusion can be drawn, namely, that the severe losses of the 1919 drought were caused principally by faulty veld and stock management.

27. The usefulness of a particular shower in relation to the vegetation depends almost entirely on the quantity which soaks into the ground.

28. A very light fall will all be evaporated and give neither run-off nor soakage. A very heavy shower precipitating water faster than the ground can absorb it will give a considerable run-off, which will increase with the intensity of the shower. But while it is evident that the usefulness or economic value of a shower depends firstly on its volume and intensity, its utility is also very largely dependent on the nature and slope of the area on which it falls.

29. Your Commission has become convinced from the evidence submitted that, as a result of the conditions created by the white civilisation in South Africa, the power of the surface of the land, as a whole, to hold up and absorb water has been diminished, that the canals by which the water reaches the sea have been multiplied and enlarged, with the result that the rain falling on the sub-continent

to-day has a lower economic value than in days past. Herein lies the secret of our "drought losses."

30. The diminished capacity of the country to hold up and utilise the rain which falls has been caused by the deterioration of its protecting vegetal cover and by soil erosion. In subsequent chapters it will be shown how this deterioration has been caused by bad veld management and how soil erosion is brought about by the same cause.

31. It is thus evident that by the improvement of the methods of farming practised in the country the efficiency of the rainfall may be increased, which, as far as the vegetation is concerned, is equivalent to an increased fall.

32. The position may be briefly outlined in the following sentences:—

> (i) **No proof was submitted that the mean annual rainfall of the Union has altered appreciably within recent historic times. Nor is it considered likely that such a change has taken place.**
>
> (ii) **According to the evidence of many witnesses there has been an alteration in the nature of our rainfall within the last few decades. No measurements have as yet been submitted either supporting or rebutting this statement; but it is well within the bounds of possibility. There is nothing to show whether this alteration, if it exists, is a permanent or only a temporary change.**
>
> (iii) **While the mean annual rainfall remains constant its economic value has, to a very great extent, been reduced by the alteration in the properties of the surface of the country for which man is responsible. In this reduced utility of rainfall must be sought the secret of our "droughts."**
>
> (iv) **In subsequent chapters it will be shown how improved farming methods and conditions will result in increased rainfall efficiency.**

III. Kraaling of Stock

33. In the first paragraph of the Terms of Reference, your Commission is instructed to investigate the possibility of reducing periodic drought losses by changes in methods of farming. As most

of the drought losses in the Union are suffered by small-stock farmers, it is necessary to discuss the methods at present practised by them in order to discover whether any such changes can be recommended.

34. Your Commission finds that the custom of kraaling or concentrating stock at certain fixed places at night is practised, with some exceptions, by most farmers in South Africa. These exceptions are cases in which sheep are allowed to run free day and night in suitable paddocks, or in parts of the country where no jackals are found.

35. Left to itself the sheep, in the hot weather, grazes during the early morning and late afternoon; he rests during the heat of the day and sleeps through the night. If he is kraaled this natural order of things is upset; he is being driven to and from the kraal, or is grazing when it is highly desirable that he should be resting in the protection of whatever shade is available.

36. Naturally, the grazing near the kraal is the first to be finished, the good grazing gradually recedes from the kraal as the winter progresses and, in a prolonged drought, have become weak as the result of insufficient food, they have further and further to travel to whatever food is available, and, probably, the time occupied in daily journeyings may be so considerable as to leave not enough in which to eat their fill; the flock is starving while there is still food enough, if only the sheep could get to it.

37. Obviously, this state of affairs could be remedied by the provision of a sufficient number of kraals, suitably located. But, if for no other reasons, owing to the greater risks which kraaled stock run from attack by internal and external parasites, and owing to the depreciated value of wool soiled by dirt collected from kraals, your Commission cannot recommend the multiplication of kraals.

38. In place of kraals, wired jackal-proof enclosures, of dimensions large enough to permit of the animals sleeping wide apart, have been recommended; and there is a good deal to say for them, as they are obviously not nearly so objectionable as kraals. Their use, however, would still necessitate the driving of sheep, the expense of herding them, and, furthermore, would be no aid whatever in exterminating the jackal.

In certain parts of the country where small stock are kraaled almost exclusively during winter nights for protection against severely cold weather, tree or shrub shelters, judiciously planted in many places on the farm, have proved to be much more satisfactory than the kraaling system. During the hot summer months such shelters afford shade and in some cases also some food for the animals throughout the year.

39. Not only does kraaling increase the difficulties of the sheep in obtaining its food, but it actually increases the food requirement, and thereby lessens its chances of coming through a drought.

40. The increased food requirement arises in the extra work it has to do in travelling to and from its grazing grounds, and, also, to a less extent, to the fact that it has to spend more energy in keeping down its body temperature (which it does by evaporating water from lungs and skin) than its more fortunate unkraaled brother who rests throughout the heat of the day. A still more important fact than any yet mentioned is the greater ability of the free ranging sheep to sustain life by drawing on the reserves of fat and flesh of its own body.

41. A sheep in a normal state of health can live several weeks without food, provided it has sufficient water to drink and provided also it is at rest. Presuming, therefore, all the grazing on the farm to have been consumed, the resting sheep (*i.e.*, the free ranging sheep) has still many days to live while his unfortunate driven brother will rapidly sink exhausted by his daily toil long before his body reserves have been fully utilised.

42. But usually the veld will not be entirely depleted of fodder although there may only be a very little. Little though it be, provided only the animal can get it, it will postpone the day when the animal will have called on its body reserves to the utmost limit; obviously the unkraaled animal has the best chance of doing so.

43. It is evident, that at the end of a bad season, the general condition of the free-ranging sheep is better than the sheep which has been kraaled, and if, as frequently happens, the following season is also unfavourable, the former animal goes into it with more reserve strength and an increased chance of coming through.

44. From whatever angle the matter is viewed, it is clear that, in addition to the expense of herding, a considerable financial loss attends the kraaling of sheep even if no deaths occur.

45. It follows from the above, that if the farmer would fight droughts successfully or would get the best return from the grazing his farm produces, he must reduce the movement of his flocks to the lowest extent possible: this he cannot do so long as he follows the system of kraaling sheep.

46. The experience of the small number of South African farmers who, having the jackal under control, have abandoned the kraaling system, proves indisputably that stock losses can be

reduced greatly or entirely prevented by the adoption of their system of allowing the stock to run free day and night.

47. Your Commission is of the opinion, therefore, that the abandonment of the system of kraaling sheep is an essential step in the reduction of drought losses.

Your Commission finds that:—

(1) **The kraaling of small stock, which forces the animal to lead an unnatural life, is the prevalent practice among farmers throughout the Union.**

(2) **The kraaling system necessitates much driving of stock and an increased food requirement, which is particularly disadvantageous in times of drought.**

(3) **Driving is detrimental to the condition of the animal and seriously endangers its life when, through the effects of a bad season, it is in a weakened state.**

(4) **Apart from its action on the sheep during times of drought, kraaling, as a general practice, is at all times detrimental to the health of the animal and the value of its wool.**

(5) **Experience has shown that the system of running sheep day and night in suitable paddocks is attended by very small drought losses.**

(6) **The abandonment of the kraaling system is a necessary step in the reduction of drought losses.**

IV. OVERSTOCKING

48. Another practice prevalent throughout the country among farmers, which has a very important bearing on drought losses, is overstocking.

49. Wherever your Commission went, witnesses stated that their farms were overstocked and that they made no special provision for the drought which they knew would overtake them sooner or later.

50. It is a problem of extreme difficulty to decide upon the number of stock a farm can carry from year to year, and this fact furnishes one very good excuse for the overstocking so generally observed.

51. The amount of grazing produced annually by a farm, *other things being equal,* depends on the rainfall. If, therefore, the rainfall and other factors remained constant from year to year, it would be a simple matter to arrive at the stock carrying capacity of any farm.

52. But the rainfall factor varies tremendously from year to year, and, furthermore, without any apparent regularity. Therefore, *even other things being equal*, it is impossible for the farmer to estimate the number of stock he should carry, if he depends on veld grazing alone. Besides, *other things* are never equal, making his difficulties so much greater. It is no exaggeration to say conditions vary so much that a farm will carry double the number of stock in some seasons that it can in others.

53. Since, up to the present, no means of predicting seasonal variations is available to the farmer, the element of speculation must enter largely into the stocking of farms, unless reserves of fodder are accumulated for bad times, which is seldom the case.

54. If the farmer, in deciding the number of stock he should hold, proceeds by the method of average, his estimate for bad seasons, made when the drought is not on him and his outlook is very optimistic, is bound to be high. Thus it is in bad years that the farmer finds himself with more stock than his farm can carry, and he suffers more or less serious loss. It is impossible to blame him for his optimism; but his failure to make provision for the drought is not to his credit.

55. The above is roughly what happens in the ordinary course, when the fluctuating rainfall and the subsequent impossibility of definitely and correctly estimating the available grazing for the coming season, is the reason of overstocking. There are, however, other reasons which have the same effect; high price of wool, an income tax and super tax calculated on the number of sheep sold and not on the value of the increase, a desire to pay off a bond in a short time, the inflated price of land, etc.

56. The position, then, when the 1919 drought came upon the country, was that farms were much overstocked and, excepting a few illuminative cases, there were no special fodder reserves. Added to this, owing to the drought being very widespread over the Union, those farmers whose foresight would have led them to send their stock away, were unable in very many cases to hire grazing. In times like these, when trekking becomes necessary, there is always a certain amount of confusion and waste of effort and valuable time on the part of farmers seeking grazing for their stock. Organisation, which would put the man who has grazing available, into ready touch with the farmer who needs it, would be very helpful in reducing drought losses.

57. The result of overstocking, long before it leads to losses by death, is over-grazing, the evils of which are cumulative and,

apparently, not generally recognised. In order, therefore, to give overstocking a proper perspective, it is necessary closely to examine the effects of overgrazing.

58. Overgrazing compels an increased movement of stock through their having, on the whole, to forage over a greater area to obtain their food requirements: this, like driving to kraal or water, tends to trample out the veld, and demands extra energy and, therefore, an increased amount of food.

59. When a farm is overgrazed continuously, it is plain that not only will the stock go into drought with the handicap of a shortage of food in prospect: they will also be in a poorer condition than if the farm were not overstocked.

60. More important than anything else is the progressive effect of overgrazing on the vegetal covering of the veld. This effect will be fully considered in a subsequent chapter.

Your Commission finds that:—

(1) **The practice of overstocking farms is very prevalent throughout the Union.**

(2) **Several causes are responsible therefor, among which are extreme seasonal variations and the optimism of the farmer.**

(3) **Animals on overstocked farms go into drought handicapped by a low condition, as well as little food in prospect, which circumstances lessen their chance of coming through the drought.**

(4) **The reserving of fodder for use in times of scarcity is a very unusual practice.**

(5) **Overstocking leads to overgrazing and all its attendant evils.**

(6) **Largely responsible for drought losses is the almost universal practice of overstocking the farm, and a failure to make any sort of provision for the drought, which the farmer knows will come on him sooner or later.**

V. Water Supply

61. Farmer witnesses were unanimously of the opinion that one of their greatest difficulties in drought time is to provide sufficient water for their stock; through the limited number of drinking places stock had often to travel long distances to water when in a

weakened condition; they believed that an adequate supply of water would often have meant live animals instead of dead ones.

62. Your Commission finds itself in full agreement with these witnesses, with this addition, namely, that too little attention is paid to the quality of the water supply, in that very often animals are allowed daily to urinate and defecate in the water they have to drink, rendering it impure. This has the effect of lowering the standard of general health, and with it, the power to withstand the hard conditions which are inseparable from droughts.

63. It is not necessary to enlarge on the driving of stock long distances to water; since, what applies to driving due to the kraaling—and this was fully discussed above—applies equally to excessive movements of stock necessitated by any other cause.

64. It is, however, to be noted that the farmer loses much of the benefit of having abandoned kraaling, if his stock have still long distances to travel to water.

65. Since it is highly desirable that there should be no misunderstanding regarding the supreme importance of an adequate supply of water for stock, it is necessary, briefly, to review the principal functions of water in animal life.

66. That an animal can live without food for several weeks is made possible by the reserves of food stored up in its body; but of reserves of water the animal system has none, or next to none. For this reason an animal cannot live many days without water.

67. Every cell of the animal body must contain its quota of water, which has to be maintained against continuous heavy losses in the dung, urine, perspiration and breath. If by reason of the animals getting too little water to drink, these vital processes rob the cells of the body of more water than they can spare, the blood begins to thicken and the temperature of the animal rises until a state similar to that of fever is produced, with a consequent rapid loss of flesh, and ultimate death.

68. Another very important point with reference to water supply is that, as the drought proceeds, the animal's water requirements increase; which is due to the fact that the dry fibrous food available in drought times, requires an increased quantity of water for its digestion and passage through the alimentary canal.

69. Unfortunately at such times the water supply is usually a dwindling quantity. It is no wonder that so many animals die, and that farmers clamour for Government assistance in the search for water in areas where great difficulty is experienced in finding it. It

is no wonder that the stock of an area are congregated where there is water to drink, regardless of food supply.

70. Your Commission finds that:—

Water is the first essential of life and the provision of adequate supplies of it is a prime necessity in fighting droughts; that Government should encourage farmers in every way possible to improve the water resources of the farm, and that improvement in this direction will act very materially in fighting drought.

VI. DETERIORATION OF VEGETAL COVERING

71. The retention or improvement of the natural vegetal covering of the country is a matter of vital importance to animal life. Obviously the vegetation valuable for grazing forms a portion of the vegetal covering, and its deterioration reduces the available food supply while the increased run-off of water, which follows upon the destruction of the vegetation, decreases the efficiency of the rainfall.

72. It is not proposed to discuss here all the causes which have led, and are still leading, to deterioration of the vegetal cover, but only certain of them, which are intimately connected with small stock grazing.

73. Kraaling and herding of stock, owing to the jackal, certainly leads to a notable destruction of the vegetal covering in a mechanical way, namely, by trampling it. Several witnesses expressed the opinion that the flocks tramp out more than they eat, which, however, is probably overstating a serious position. The most serious damage is done near the kraal where the trampling is most concentrated, and, as will be shown later, the continuous concentration of the flocks in the neighbourhood of the kraal leads to a considerable area around it being overgrazed, with disastrous results.

74. The herding of stock also tends to a similar mechanical destruction of the herbage, for herded stock keep much closer together than stock which have a free range.

75. In how far the collection of much of the dung and urine of the animals in the kraal affects the nutrition of the vegetal covering, it is not possible to say; but, presumably, it would be best if all the excrement of the animals became available as food for plants on which they fed, as would be the case if the stock were not kraaled.

76. The scarcity of drinking places causes a concentration of stock with similar evil results.

77. Speaking broadly, overgrazing modifies and, in bad cases, destroys the vegetal covering.

78. The chief characteristic of the indigenous covering of an area of low rainfall, or an area where the rain falls at infrequent intervals, is the high proportion of perennials.

79. When a rain has fallen, the perennial plant, with its already well developed root system, is in a position to make the fullest use of it, and is in vigorous growth before the annual has had time to germinate, let alone establish a root system. Thus it happens that the annual has a poor chance of surviving, unless rain falls in such a manner as to keep the soil sufficiently moist to make growth continuous. Supposing the annual makes a start and no more rain falls: it must die, but not so the perennial which, having stored up nutriment in its buds and root stock, goes into a resting condition until the next rain comes.

80. Thus, not only are perennials well fitted to make the best of a scanty rainfall and to survive long intervals between rainy periods; they tend also to repress annual growths, and in doing so mask their existence.

81. Suppose in such an area the perennials have been rooted out: the annuals will have a clear field; will have all the rainfall to their own use, will become increasingly prominent; but will still, however, cease growing and will die when the soil dries out, which, it is true, may not happen until considerable growth has been made. It is clear, however, that yield of fodder by annuals is much more dependent on rain at frequent intervals, than is the yield of perennials. That being so, any factor which tends to the replacement of perennials by annuals tends to make the veld less certain of being able to carry its quota of stock throughout the year.

82. The effect of overgrazing is actually to weaken the perennials, and thereby increase the hold of annuals, which it does in the following way:—

83. The continued existence of a perennial is only possible so long as it can produce and retain, for a sufficient period, a leaf surface, to manufacture food for storage for the resting period between rains and the period of renewed growth after rain.

84. But, if a perennial be allowed to form leaves which it retains for too short a period, it will have cost the plant more in stored food to produce the new growth, than that growth will have enabled it to re-store. If this happens without intermission for a sufficient

number of times, the food reserves of the perennial will become entirely exhausted and, like the annual, it will die. Something like this happens on overgrazed veld. The perennials spring into edible growth first, and before they have had time to manufacture their reserve food requirements, the hungry animals come along and eat them down. Thus occurs the first weakening which, repeated often enough, kills the plant.

85. Palatability plays an important part in actual practice, with the result that those perennials which the animals like best, are the first to suffer, while those they do not like at all, become more and more vigorous. Thus, in such of the drier parts of the Union where overstocking has been pronounced, annuals like "steek" grass have become prominent, and "bitter bos" has crowded out almost everything else.

86. What, then, in the neighbourhood of kraals and drinking places, is usually ascribed solely to the trampling effect of animals, has really, to a great extent been caused by overgrazing; for there, if there be any vegetation at all, it is usually of the undesirable type.

87. The effect of overgrazing on the covering of the veld is modified considerably according to the time it takes place, the most detrimental period being that when growth is very active, as for example, after the drought breaks; for this rapid growth after plenteous rains determines, not only how much fodder there will be for the coming dry period, but also, and even more important, the amount of storage of rootstock, bud and seed; and through this, the yield of fodder in the coming year.

88. Thorough grazing subsequent to this period does not seem to matter much, for from America, where important investigations have already been conducted into range management, it is reported that, by reducing the number of stock during the main growing season (about 4 months) to about half the average number the range can carry for the year, and thereafter grazing fully for the remaining eight months; it was found that the range so treated improved as much as similar ranges protected for the whole year.

89. This is most important testimony in favour of dividing farms into small camps so as to allow of proper grazing control; and it is eloquent corroboration of the evidence of those witnesses who told your Commission how the internal fencing of their farms had so improved their veld as to enable them to carry from 33 per cent to 75 per cent more stock. Furthermore, it strongly supports the reserving of fodder for drought times (and for the winter

period) for, the better the condition of animals when new growth begins, the less intensively will they graze it.

90. This explanation of overgrazing would not be complete without reference to that valuable Karroo fodder plant, pentzia virgata, the "skaap bos" and a method by which it propagates itself if given the chance. The branchlets of this plant grow over in graceful curves until they touch the ground where they bend suddenly upwards. At this bend roots form under favourable circumstances, and with them an independent plant comes into being. If stock are continually passing over these plants, it is obvious that these branchlets cannot strike root, even if they escape being eaten. Therefore, to maintain a good covering of this valuable fodder plant, the veld must be rested from time to time; nothing is more calculated to destroy it than overgrazing at the period of active growth.

91. It has been necessary, somewhat fully, to treat of the effects of overstocking, not only to show that the opinions of witnesses regarding the evil of overstocking are well founded; but also to show that the evil is much deeper seated, and more insidious in its attack than is generally understood; furthermore, it is only by taking all of the facts outlined above into consideration, *that the foundation of a proper system of veld management is obtained, namely, complete grazing control.*

92. For a farmer to have complete control of his grazing, means that the farm must be divided into many camps so that, at discretion, stock may be entirely excluded from any given areas, or be made, intensively, to graze others. The reason for exclusion will be apparent; but perhaps it is necessary to add a few words regarding intensive grazing. If intensive grazing be not practised, stock will confine their choice to the most palatable plants. This will eventually lead to their destruction and the spread of the less palatable growths. With intensive grazing stock are driven by hunger to eat all growths which are not actually distasteful, so that all useful growths get an equal chance, when the camp in the course of rotation, is rested. Plants not eaten, or which have become established under such conditions, will early become manifest, and steps can be taken to eradicate them before they spread and crowd out useful growths. It may also be stated that palatability and nutritive quality do not always go together; in fact the less palatable growths are often the most nutritious; an additional reason for compelling stock to eat them.

93. No matter how carefully grazing is handled, numbers of valuable plants are bound to die every year, tending to reduce the density of the vegetal covering. For this reason parts of the veld should, from time to time, be allowed to seed before they are grazed. With the farm divided into camps, it will be an easy matter to arrange this.

94. Your Commission is of the opinion that, in order to prevent the deterioration of the vegetal covering of the veld, as well as to improve deteriorated veld, it is very necessary to prevent its trampling by stock and, in particular, the overgrazing at the period of active growth following the breaking of the drought.

95. The deterioration in the vegetal covering, which is closely connected with the operations of the small stock farmer in the drier parts of the Union, has been shown to be due mainly to kraaling, scarcity of drinking places and overgrazing.

96. Veld burning, also an important factor in the case, has been omitted in this report, as sufficient information has not yet been collected, and because veld burning is not practised in those parts of the Union where stock losses are severest.

97. In addition to what may be called the external causes of deterioration, another cause is found in the deterioration itself, so that, where other contributory factors remain unchanged, the deterioration, as it increases, will, within certain limits, proceed with accelerated speed. It has already been pointed out that deterioration, by forcing additional movement of stock, increases both overgrazing and the mechanical damage to the vegetal cover; it is necessary now to introduce the relation between reduction of vegetal cover, and efficiency of rainfall, in order to show how the reduction of either tends to a reduction of the other.

98. The first and obvious point is accelerated run-off following on a reduction of cover. It is unnecessary here to discuss in detail how this accelerated run-off is caused, as the fact is self-evident. This increased run-off which is attended by a decreased rainfall efficiency is serious to vegetation growing under conditions where the water supply is the limiting factor.

99. As serious as the increased run-off in reducing the rainfall efficiency, is the increased evaporation, occurring as a result of deterioration of cover.

100. Rains in the drier parts of the Union give place very suddenly to fine hot weather, so that the evaporation of the rain begins very soon after it has fallen; often long before it has sunk well into the soil. Now when a soil surface is wet the rate of

evaporation from it in hot weather is very considerable, thus a large percentage of the rainfall never becomes available to plant life. The effect of the destruction of the vegetal covering, by giving a greater exposure of the soil surface to the sun increases the rate of evaporation and, through that, the proportion of the rainfall lost to plant life.

101. Where the extent of the reduction of cover is serious, the loss of moisture due to the increased run-off and evaporation may become so great that the total amount available will be insufficient to support the original vegetation. When such a condition is reached, rapid deterioration results, unless other contributory factors, which man may control, are made more favourable to recovery. This is another argument for improved veld and stock management; because, the rainfall which becomes available to plant life is the principal factor in the yield of grazing.

102. Your Commission finds that:

(1) The kraaling and herding of stock leads to a mechanical destruction of the vegetal covering due to trampling.

(2) The lack of a sufficient number of drinking places gives rise to a similar result.

(3) Overstocking not only leads to trampling, but also to overgrazing.

(4) Overgrazing tends to destroy perennial fodder plants and encourages the growths of annuals and plants useless for grazing purposes. In this way the grazing yield of a season is diminished and depends more and more on frequent rains.

(5) The effect of overgrazing is very serious when it occurs during the main growing season.

(6) The farmer should therefore endeavour to reduce intensive grazing at this period.

(7) This he can do if his farm is divided into paddocks, for such a sub-division permits of the best possible distribution of the stock over the farm, and allows of absolute rest for paddocks which require it.

(8) Complete grazing control is the first essential of a system of stock farming that will prevent deterioration of the vegetal covering.

(9) Animals in poor condition graze more destructively than if in good condition.

(10) Reserves of fodder for use, when grazing is scarce, are very valuable, not only for keeping stock alive, but also for

preventing overgrazing at the critical time when vegetative
growth is very active.

(11) Even if no permanent damage is done, overgrazing at the
period of active growth seriously diminishes the following yield of fodder.

(12) Deterioration in the vegetal covering of the drier parts of the
Union has been brought about, mainly through the practices of kraaling, herding and overstocking, together with an
insufficient number of drinking places, and overgrazing.

VII. Soil Erosion

103. The Report up to the present point has been concerned
mainly with an examination of farming methods and conditions
in their relation to periodic drought losses, and it will be found, on
further examination, that these same factors, which go to cause
drought losses, also cause soil erosion, which will now be discussed.

104. It is necessary, before entering into a discussion as to how
the prevention of soil erosion may reduce drought losses, to outline
how soil erosion has been brought about, and the damage it is
doing.

105. Soil erosion can be separated into two divisions, the erosion
of cultivated lands, which is a matter of extreme importance,
although, until the present time, little or no investigation of this
problem has been made in South Africa, and veld erosion as dealt
with under Par. 106. The area thus far covered by your Commission in its investigation has not included any portions in which
erosion of cultivated lands is very serious, and, for this reason only,
and not on account of its being of lesser importance, your Commission has decided not to deal with it here.

106. Your Commission finds that the soil of South Africa is
being eroded rapidly in three ways:—

(a) By surface erosion by wind.
(b) By surface erosion by water.
(c) By donga or sloot formation.

107. The surface erosion is, in a sense, the more dangerous form.
It is an insidious evil creeping in unseen like a thief in the night
and robbing us of our national wealth. The far-reaching effect of
this evil is not yet realised, even by many of the soil conservation
enthusiasts in the Union.

108. Your Commission finds the great important truth, that *the*

soil of the Union is a definitely limited and irreplaceable quantity,
is not generally realised. Although more soil is continually being
formed, the time taken to produce even so small a quantity as an
inch depth of soil, is so long in comparison with the span of
human life, that the little formed during a generation may be
neglected as far as we are concerned.

109. Since then the soil is a definitely limited quantity, we are
morally and economically bound to conserve it. It is the greatest
national asset, the ultimate source of all food, animal as well as
vegetable, the source of all clothing, furniture and a large part of
our buildings! Slooting, unlike surface erosion is always visible
and more or less evident from its first beginnings as a little
uncontrolled stream hurrying on its tortuous course. It is something which can be easily realised and is patent to all. Surface
erosion, on the other hand, frequently takes place without its being
evident and in the continual removal, with greater or less speed, of
the surface layers, the country is losing the most valuable soil and
plant food. No figures are available on which an estimate of the
plant food annually lost in the Union may be made, but the loss
must be enormous.

110. The wind is similar in its action, to water. It is the rich
surface soil, which has taken centuries to form from the sub-soil,
which is first removed. In certain portions of the Union ploughed
land has been bodily removed and piled in huge sand dunes which,
in their progressive rolling, suffocate everything in their path, and
leave only blackened roots in their wake. But this very visible type
of wind erosion causes only a small portion of the total damage
inflicted. The carrying capacity of the wind is enormous, and when
one compares the number of hours per year, during which a
scouring sheet of water is passing over the more arid portions of the
Union with the number of hours that a strong dust-bearing wind is
blowing over the same area, the relation between wind and water
surface erosion becomes more evident.

111. While water can carry the eroded material in only one
direction—seawards—wind may carry dust uphill or down dale.
And yet ultimately this dust, too, moves in only one direction, for
however much it may be tossed about backwards and forwards, the
direction of the wind prevailing during the dry season must and
does decide the main resultant direction of motion.

112. In addition to this surface erosion, and going on simultaneously therewith, indeed greatly assisted by it, is the slooting of our
country; that is, the cutting up of the veld by runlets and gulleys

which eventually form the deep water courses we know as sloots or dongas. These sloots vary from a few inches to many dozens of feet in depth, depending upon the volume of water passing down them, the nature and depth of soil and sub-soil, the gradient of the original surface, and the presence of dykes of rock or similar controlling factors.

113. The damage done by surface erosion is the removal of valuable surface soil. Sloot erosion removes both soil and water. The entire soil formerly covering the area occupied by a sloot has been removed to a depth of perhaps dozens of feet. It is true that the sub-surface soil may not be so valuable as that which lies above it, but it acted as a source of plant food and a reservoir for water.

114. In addition, by the reduction of the level of the water surface in the deepening watercourse or sloot, the virtual gradient of the surface water on the banks of the sloot, during a rain, is greatly increased. This results, in the first instance, in innumerable branch sloots eating their way back from the banks in ever-increasing ramifications and, consequently, in an accelerated surface erosion in the surrounding area. In this manner *all* the surface soil is eventually removed in the vicinity of sloots in many portions of the Union. The resultant bareness produces an increased run-off, that is, a reduction in the quantity of water soaking into the ground. The deep sides of the sloot further tend to drain the soil on its banks of the water which has managed to soak in.

115. The increased run-off, which all finds its way into the sloots, enormously increases the power of water to undercut the banks, and transport the large amount of material which has caved in during the dry season.

116. Surface erosion, by causing more rapid accumulation of water, multiplies the number, and increases the size, of existing sloots. This, in its turn as shown above, accelerates surface erosion in the vicinity and, by the removal of valuable soil and water, further increases the devastation and prevents recovery.

117. *The cumulative character of soil erosion, noticeable in all its phases, is its very worst feature and supplies an incontrovertable argument for immediate and prompt action if retrogression is to be arrested.*

118. Water, which should have soaked into the ground to feed plants and replenish the underground supply, is carried to the sea. Your Commission has found that while, generally speaking, there was a considerable sinking in the water table or depth of the underground water throughout the area traversed, the water, to a

notable extent, recovered its original strength during recent wet years, except in those districts where erosion is particularly bad.

119. In cases where the source of the water supply can be traced, the relation between rainfall, soil erosion and depth of ground water is quite clear. Springs which derive their water from a long distance reflect variations in the rainfall only faintly, and with a considerable period of retardation between cause and effect. Local shallow springs are affected almost immediately by local rains, while in all cases it appears that an eroded catchment, badly slooted, spells a continually receding average water table, however much it may oscillate in answer to changes in rainfall.

120. The sinking of the water table is an economic loss to the country—a reduction in the value of our assets, in so far that greater energy is needed to bring it to the surface where it can be used, and that the ground water is liable to be less available to crops.

121. These ever increasing sloots must result not only in an increased but also in an accelerated run-off. Owing to the recent date of our river gaugings, this point cannot yet be established by actual measurements, but there can be no doubt on the matter. As a result floods in our rivers may be expected to increase in severity, but decrease in time of flow, while periods of no flow will naturally become proportionately longer.

122. South Africa is, to a large extent, an arid or semi-arid country and needs a great deal of irrigation, if for no other purpose than for the full development of its cattle raising potentialities. One stumbling block in the way of irrigation extension is the irregular flow of our rivers. The high floods necessitate exceptionally costly protective works: the short floods abnormally large canals to permit of the flow of a sufficiency of water in the short time available. These two points, combined, frequently make irrigation schemes economically impossible, the ever-increasing erosion, by increasing the irregularity of our rivers, is magnifying the difficulty and the production of feeding stuffs is hindered.

123. The remedy usually recommended is to build reservoirs for the purpose of regulating the flow to a certain degree. Unfortunately, recent experience in many parts of South Africa has proved the silt factor to be serious. By reducing the useful life of the reservoir the silt increases the amount of repayments to a sinking fund and thereby the annual cost. More erosion naturally means more silt and, therefore, a higher cost of production of feeding stuffs.

124. While some of the silt carried down our rivers may be beneficially used again, as in the Lower Orange River, where the crops are grown on alluvial soil brought down from the uplands, or as along the Gamtoos River, where the silt of the Groot River rejuvenates the old lands, yet the greater portion of the silt finds its way, unused, to the sea. And even the silt which does not find its way to the sea is not always beneficial in its results on vegetation, for in certain cases the very fine silt chokes or suffocates plant growth so that its presence in water renders it unfit for irrigation; while in many other cases, again, coarse material brought down from upper slopes spreads out, fan-like, on reaching gentler slopes and kills off all vegetation except useless weeds, thus reducing the grazing yield of the area.

125. The erosive power of water is enormously increased by concentration and increased velocity, and it is to the factors that bring about concentration and increased velocity that we must look when searching for the causes of soil erosion. Generally speaking, the increased velocity is due to a laying bare of the surface of the country affected; but the actual extent of the soil erosion effected depends on many factors, such, for example, as climate, including the temperature, humidity of the air, as well as annual rainfall and the maximum intensity thereof. The resistance of a given soil to erosion depends also on the physical and, in special cases, on the chemical composition of the soil and the nature of the sub-soil and, naturally, also on the slope of the surface and the amount of run-off reaching it from higher-lying ground. The speed at which soil erodes is also decided, to some extent, by its aspect with reference to the direction of prevailing winds and rain storms. And last but not least a big controlling factor is the amount of vegetable covering by which the soil is protected. For a given climate, soil, aspect and slope, there is a certain minimum of vegetal cover needed to prevent erosion. *When left to herself Nature arranges a state of balance between the various factors, rainfall, run-off, soil, aspect, slope and vegetation suitable to the climate. When Man arrives in the arena and upsets the balance by destruction of the vegetation, trouble must result.*

126. While man has been responsible for much destruction of vegetation in the Union, probably no one class of destruction has affected so large an area as that occasioned by the small stock farmer in the manner detailed in preceding chapters. The wasteful system of veld management, which causes so much deterioration of

the vegetal cover, therefore can be seen to be a fruitful cause of soil erosion.

127. Indeed, it is now evident that *the present system of grazing is detrimental to stock vegetation and soil.*

128. While this is a very unsatisfactory state of affairs from one point of view, it is fortunate that as a result of the interdependence of these factors, one remedy, viz., improved methods of veld management, will simultaneously reduce the evil effects of all.

129. Improved veld management is, therefore, much to be desired, and although in the Terms of Reference your Commission has been instructed to report, mainly on direct stock losses, it is deemed advisable to point out here that, while the loss of stock leaves no permanent after-effect, both deterioration of the veld, through retrogression of the vegetal cover and soil erosion, are far-reaching in their effect, and their results will long be felt.

130. For this reason the adoption of a better system of veld management is as necessary for the welfare of future generations in the Union, as for the saving of the flocks and herds now grazing on our veld.

131. It is convenient here to draw attention to the dependence of other phases of soil erosion upon veld management.

132. Kraaled and herded sheep, for instance, do unnecessary mechanical damage to the soil, just as they do to the vegetation by kicking it loose with their hoofs and thus facilitating wind erosion.

133. Then, again, while your Commission does not desire to enter into the whole question of the effects of roads and railways on erosion at the present stage, it is well to mention a few points. Roads and railways undoubtedly interfere with the natural flow of the water and, by so doing, bring about a dangerous concentration, yet had the surrounding veld not been despoiled by wasteful farming, the run-off would be less, the concentration by roads and railways less and, therefore, the damage done thereby less. Thus, any successful efforts of the State to reduce deterioration of the veld will, *ipso facto*, reduce the erosion placed to the debit of roads and railways.

134. With reference to soil erosion, your Commission has found:—

(1) **That soil erosion is extending rapidly over many parts of the Union.**

(2) **That, besides slooting, there is a great deal of surface**

erosion, both by water and wind, taking place.

(3) That the soil of the Union, our most valuable asset, irreplaceable and definitely limited in amount, is being removed in enormous quantities annually.

(4) That the greater part of this soil and valuable plant food is lost for ever, and while the remainder of the eroded material may do good in some instances, it does much harm in others.

(5) That one great damage done by the eroded material is in silting up of reservoirs and that soil erosion causes greater irregularity in the flow of our rivers, thereby increasing the cost of irrigation works and the cost of producing feeding stuffs.

(6) That soil erosion is causing a marked decrease in the underground water supply of the Union, and thereby increases the difficulty of watering stock.

(7) Soil erosion is caused by reduction of the vegetal cover.

(8) That soil erosion has a cumulative character which, by virtue of the similarity between its cause and effect, always accelerates its rate of growth, in all except a few favoured portions of the Union.

(9) That prompt action is therefore imperative.

(10) That soil erosion is caused, mainly, by deterioration of the vegetal cover, brought about by incorrect veld management, and that all efforts to improve the latter will have a beneficial result on the former.

VIII. Improvement of Farming Methods and Conditions

135. Sufficient information has now been submitted to make it perfectly clear that the present system of small stock farming, with its kraaling, herding, overstocking and uncontrolled grazing, is leading to serious damage to both soil and grazing.

136. So important to the country is the abandoning of this system and the general adoption of paddocking and improved system of veld management that your Commission is making a survey of the grazing potentialities of the Union with a view to estimating what would be the increased carrying capacity of the country if paddocking be adopted. These figures are unfortunately not ready for inclusion in this report.

137. Evidence showed that, by the adoption of the principle of free ranging an increase of 75 per cent of stock was carried on a

certain farm without damage to the veld, which, on the contrary, actually improved. Similar evidence was received also from other farmers. There can thus be no doubt as to the success of the proposed system, yet with few exceptions, it has not been adopted generally throughout South Africa. Before discussing recommendations for putting into effect those remedies for drought losses which have been discussed in the foregoing pages, it is advisable to endeavour to ascertain the reasons for the retention of the kraaling system.

138. Your Commission, after sifting of the evidence, finds that the present system of small stock farming is being adhered to an account of:—

(a) The presence of the jackal, which necessitates kraaling.

(b) The scarcity of natural water supply for the drinking places which must be provided in every camp if paddocking is adopted.

(c) The want of capital required to erect the necessary jackal proof and other fencing, and to provide water for the paddocks.

(d) The presence of roads—many of them unnecessary—which make the lay-out of a suitable scheme of paddocking extremely difficult, or indeed impossible.

(e) Custom and the lack of a full realisation, on the one hand, of the evil results of the veld deterioration and soil erosion caused by present methods and, on the other, of the advantages of the new system.

139. It is evident to your Commission from the evidence and from a study of the conditions in the Union that while certain individuals are fully cognisant of the evil which must result from this effort, they start farming with the set purpose of wringing out the life blood of the farm in order to make a quick profit; the great majority allow their farms to be damaged through a failure to recognise the danger on the one hand or to perceive clearly the causes on the other.

140. Many witnesses have most strongly recommended direct legislation, stringently administered, with a view to stopping the various evils which have been presented in this report. It is therefore necessary to give the findings of your Commission in this connection, and the discussion will at the same time serve to show the limits of public and private action in the reduction of drought losses.

141. First and foremost the State is bound to prevent such waste

of its natural resources, which, if persisted in, can only but lead to national suicide; so it should take action in connection with soil erosion as it has done in other directions.

142. The State has already recognised the national importance of river waters and has enacted far-reaching and profound legislation thereon. This principle, carried to its logical conclusion, would seem to demand State control of the factor of run-off. Large problems of this nature are looming in the not very distant future.

143. While the responsibility of the State in this matter is great, the individual also has his responsibilities. The individual has brought about the damage and without his co-operation the damage cannot be repaired.

144. Prevention and sustained vigilance are the watchwords necessary in this work. No State organisation can ever supply the minute watchfulness needed, and can but look to the individual to undertake the work for the good of himself, of his children, and of the State.

145. Although it may be highly desirable to give effect to the recommendations of those witnesses who desire direct legislation, your Commission does not consider the time ripe for such action. Education of public opinion is first required and thereafter direct legislation, if still necessary.

146. Your Commission therefore recommends the organisation of a sustained campaign of propaganda in connection with veld deterioration and soil erosion in order to bring home the dangers thereof to all: failing this, there would seem to be no prospect of any appreciable change for the better.

147. As the result of travelling through the country and coming into close contact with the people, your Commission has come to the conclusion that the Department of Agriculture is not in that close touch with the farming community which is so highly desirable, especially for propaganda purposes.

148. While the reasons for this lack of touch were not investigated fully, it was evident that largely responsible is an insufficient personnel, especially in regard to the Sheep and Wool Division.

149. Farmers complained that there were far too few sheep and wool experts to meet their needs. Since the best returns from sheep farming are only possible when the animals lead a natural life, your Commission is of the opinion that the propaganda natural to the calling of the sheep and wool experts is bound to be attended by far-reaching results in the direction of improved veld management and the prevention of soil erosion throughout the Union, provided

the number of such experts is adequate.

150. Therefore, in supplying the existing need for more sheep and wool experts, Government would at the same time be doing a good deal in the direction of preventing the spoliation of the veld.

151. Your Commission has come to the conclusion that instruction on soil conservation should occupy a place in the curriculum of every educational institution in the country.

152. To the Orange Free State goes the credit of having been the first to introduce the study of soil erosion to the school child, and her sister province, Natal, followed suit a few years later; but the two larger Provinces of the Union have, as yet, made no move in this laudable direction.

153. Your Commission finds that:—

(1) **The retention of the old wasteful and destructive method of kraaling and herding small stock is due to several reasons, the chief of which is the presence of the jackal.**

(2) **Other reasons are scarcity of water, want of capital and the presence of roads.**

(3) **Lack of full realization of the advantages and disadvantages of the two methods of small stock farming also plays an important part.**

(4) **Educative work is now highly necessary to induce the individual to do his share.**

(5) **The State has grave responsibilities in preventing the waste of the natural resources; but direct legislation cannot now be recommended.**

IX. INDIGENCY ARISING OUT OF DROUGHT LOSSES

154. Before coming to the recommendations for the remedying of the evils already discussed, it is necessary to introduce the subject covered by paragraph 3 of your Commission's Terms of Reference. The number of indigents drifting from the countryside to the cities appears to be increasing and this very serious state of affairs demands investigation. At the very commencement of this chapter your Commission desires to draw attention to the fact that the indigent, or poor white rejected by the land, is only the extreme cause of failure.

155. The path of the farmer in South Africa is beset with many obstacles and set backs which, while they only hamper the progress of the capable, present increasing difficulties to the less capable

members of the profession and may utterly break the weakling.

156. Indigency is due, not only to economic reasons, but also to a psychological and physical state, brought about by such causes as unfit parents, inbreeding, underfeeding, disease and climatic conditions. Among other reasons for the failure of farmers to retain possession of their land in South Africa the following may be mentioned:—

(*a*) Droughts, hailstorms and untimely frosts.

(*b*) Jackals.

(*c*) Stock and plant diseases and insect pests working separately or in conjunction with droughts.

(*d*) Cataclysms, such as war, rebellion, strikes, "slumps," etc.

(*e*) Too minute sub-divisions of farms.

(*f*) Inflated prices of ground.

(*g*) Want of general agricultural education and training.

(*h*) Inability to dispose of farm produce at reasonable rates.

157. While the instructions to your Commission, as set out in Section iii of the Terms of Reference, are confined to the indigency arising out of drought losses, the nature of the problem is such that it is impossible to separate one small portion of it from the whole interlaced mass of interdependent and simultaneous causes of indigency.

158. Many of these causes work slowly and are not very noticeable until a drought, by making surrounding conditions more severe, brings on a climax.

159. Your Commission is not yet in a position to deal with indigency, due to drought losses in its entirety. An endeavour is being made, as a commencement, to trace the movement of the population and thereby to determine, if possible, to what extent drought is the direct cause of indigency. But whatever the results of these investigations may be, one fact must remain, and that fact your Commission wishes to emphasise, namely, the improvement of any of the conditions in connection with the causes enumerated above must result in a reduction of the failures at all times, and particularly at times of sudden stress, such as drought periods.

160. Building upon this truth, your Commission very early in its investigations decided on the urgent necessity for improving the state of affairs as far as the two last terms of the list of causes of indigency, mentioned above, are concerned, namely, training and marketing.

161. Increased facilities for marketing would undoubtedly reduce poverty on farms, and where the produce is of a perishable nature, factories for transforming them into non-perishable commodities are desirable.

162. Your Commission does not suggest that at the present time the State necessarily should erect factories for, say, condensed milk, or cold storage depots for the preservation of fat carcases; but such things are urgently needed in the Union, and private capital would perhaps be available if the supply of raw material were assured.

163. It is frequently stated that farmers are spoon-fed. Undoubtedly the farmers, as a class, are very prone to leave a great many matters in the hands of the Government. The Government is continually being called upon by the farmer to perform certain duties which are not a true function of the State at all. In many instances, this amounts merely to a loan of the State organisation to the farming community by the Government, for the performance of some function which the farmers could have carried out equally well had they been in possession of the necessary *organisation*.

164. If the agricultural producer be properly organised, capital would find easy contact with the farmer, and the establishment of all sorts of factories would be much facilitated. When such a condition is reached, the farming community becomes more self-reliant and the responsibility of the State is decreased. The State organisation is then no longer called upon to carry out extraneous duties, and may be curtailed. Organisation of agriculturists (in the widest sense, including stock farmers) will thus be seen to tend to improved markets, the establishment of factories to fewer poor whites, to increased revenue and, perhaps, to less costly Government.

165. But another one of the factors, causing failure on farms which has been mentioned, namely, the question of want of agricultural education and training, also stands in very close relation to the question of organisation. It has been stated as one of the findings of your Commission that the efficiency of the technical staff of the Agricultural Department is hampered by a lack of contact with the farmer, and that the bridging of this gap by a link of some nature is necessary.

166. In the organisation of the farmer, is to be found the link.

167. Your Commission finds:—

(i) That the frequent failure of the farmer in South Africa is due to many causes, which so frequently work simultaneously, that it is difficult to separate them. Periods of sudden strain, such as drought or economic cataclysms, accelerate and magnify losses due, in the first place, to other causes.

(ii) That, if the pressure from any of these causes be reduced, losses due to droughts will also be diminished. The two causes, mainly dealt with in this chapter, are lack of technical advice and lack of suitable markets.

(iii) That organisation of the farming community will tend to improve the marketing facilities for all sorts of produce, and will form a link between the Department of Agriculture and the farmer, which will be particularly useful in spreading information.

(iv) That this will result in increased production and fewer failures among the farming profession.

X. RECOMMENDATIONS

168. If the conclusions arrived at by your Commission are well founded—and it would seem they are indisputably so for they rest on the solid support of practical experience corroborated by scientific analysis—it is obvious that most of the losses that have occurred through drought would have been prevented had the paddock system of veld and stock management generally obtained.

169. Your Commission therefore recommends that the Government should do its utmost to abolish the kraaling system and make it as easy as possible for the farmer to put the paddock system into effect.

170. In order to make this change possible, it will be necessary, as has been shown, to exterminate the jackal, to provide fencing on easy terms and to facilitate the provision of drinking water for stock.

171. It has also been shown that the deterioration of the veld and soil erosion are national dangers, and that the State must assume its responsibilities in connection therewith.

172. In dealing with all these matters and in order, generally, to improve farming conditions, organisation of farmers is a first essential.

173. Although your Commission, throughout the report, has pointed out the great and certain advantages attached to the paddock system of stock farming, it is of the opinion that the best

results from this system, for various types of veld, will only be obtained after a thorough investigation into the many problems connected with the grazing of stock.

174. A large amount of evidence was submitted in praise of ensilage, prickly pear, American Aloe (Agave) and the "norsdoring," and your Commission has reason to believe that this praise is well founded, but is of the opinion that further information on these matters is needed.

175. The essentials recommended by your Commission, as far as the matters of this Report go, are:—

   (i) **The organisation of the farming community.**
  (ii) **The extermination of the jackal.**
 (iii) **The provision of cheap fencing material.**
  (iv) **The development of water supply for stock.**
   (v) **The adoption by the State of its responsibilities in the control of soil erosion.**
  (vi) **The investigation by the Agricultural Department of certain grazing and fodder problems.**

176. **Your Commission, fully alive to the necessity for present-day strict economy, presents only these recommendations for your earnest consideration because they all deal with those aspects of the drought problem which are cumulative in their character heaping up trouble for the future at an increasing rate, and demand therefore immediate action.**

XI. Organisation of Farming Community

177. The organisation of the farming community has been touched on in the chapter dealing with Indigency, and several points were mentioned in approaching the question of organisation as a panacea for the evils of lack of training and insufficient markets. It is through these avenues of thought that your Commission arrived at the consideration of organisation as an aid to the solution of the problems summed up in the Terms of Reference.

178. There are, however, many other arguments in favour of organisation. From the point of view of the Government of the country, for example, organised farmers are easier to deal with as their wants and desires are more easily ascertained. Organised farmers are also more independent and self reliant, and do not fall back on the Government for support and assistance as do unorganised farmers. Some years ago the Western Province Wine Farmers

held demonstrations in Cape Town; to-day, now that they are well organised, they ask no assistance from the Government, but by organisation and co-operation, they have solved the problem of supply and demand in an elementary and perfectly effective manner, and have kept the price of their produce at a reasonable figure. This action has relieved the Government of much worry. *It is in the interest of the State to have the farmers organised.*

179. The utilization of farmers' organisations, in arranging for the required contract between the Department of Agriculture and the farmer, has already been mentioned. Such organisations not only ensure audiences for the specialist on his visits, but when several neighbouring organisations work together, long and useful tours, with a minimum waste of time, may be arranged. This all tends to increase the usefulness of the Department at a reduced cost, and assists in the spreading of that knowledge and experience collected by the specialist at the cost of the State, with the result that losses of all sorts are diminished, and production increased. Even the ordinary meetings of the organisation, by promoting the exchange of ideas among farmers, have a high educative value exclusive of the information imparted by the specialist.

180. The effect of farmers' organisations on the enticing of Capital for the purpose of converting the raw material into manufactured goods has been mentioned. Farmers' organisations also tend to stiffen up local prices, for by the publishing of market reports among their members, the organisation reduces the chances of the "sharp" middleman. They also reduce the opportunity for many kinds of roguery practiced on the farmer, such as the sale of shares in bogus companies dealing with agricultural matters, sale of badly bred animals for stud purposes, etc.

181. Organisation in certain parts of the Union, resulting from propaganda work carried out by your Commission, has been reported successful in staying financial panics, which would have resulted in bankruptcies and poor whites.

182. For all these reasons organisation of the farmers is necessary, and the sooner they are organised the greater the advantage to the State. With the object of achieving all these benefits and, ultimately, being freed from unnecessary work, the State should do all in its power to bring about organisation.

183. If the matter were left entirely without intervention, all the farmers, throughout the Union, would eventually organise themselves, as some of them already have in certain districts, but the time that would elapse before its consummation is longer than we

can afford to wait. The history of our country during the past decade has unfortunately left behind a too well defined political cleavage in the population, and this is a sufficient reason why the local leaders find organisation of the whole farming body so difficult. Generally speaking, the best local men, who could best undertake the organisation, are all men who have taken a leading part in politics on one side or the other, and to-day carry so marked a political atmosphere with them that their political opponents view them with distrust, even in agricultural matters. Organisation, therefore, to have a chance of success, must be undertaken from outside by someone, such as a civil servant, who has no politics, or some other man whose political preferences are unknown.

184. It is unnecessary, in this report, to discuss, at great length, details and methods which should be employed to carry out this organisation. A small staff of bilingual men, with thorough understanding of the South African farmer and agricultural conditions, will be necessary with the assistance, from time to time, of some farmers from another district.

185. **Your Commission recommends that the Department of Agriculture immediately proceed with the organisation of the farming community.**

XII. Extermination of the Jackal

186. In the foregoing pages the practice of kraaling and herding stock was shown to be necessary on account of the depreciations of the jackal; also, that the abandonment of this system of small stock farming for a system, under which the sheep can live a natural life, is the first essential in preventing drought losses, deterioration of the veld and soil erosion.

187. The logical conclusion to these findings is that the jackal (and other carnivora which attack sheep) must be brought under control and, if possible, exterminated.

188. The natural importance of jackal extermination is already recognised by the State in the rewards paid for his destruction, and in the loans which farmers can obtain from the Land Bank for the erection of jackal-proof fences on their boundaries; but your Commission is of the opinion that the State should do still more.

189. In this connection the question of increased rewards has been considered. While there is a large body of opinion in favour of

increased rewards, which it is said would lead to professional vermin killers, there is also considerable opposition to such increases. Your Commission makes no recommendation on this point.

190. According to the evidence placed before your Commission it is, however, plain that little permanent success will attend the efforts to destroy the jackal in any district unless he be first pinned down to an area which will permit of his being eradicated rather than driven away.

191. While there is a considerable difference of opinion throughout the country as to the best method by which the jackal can be exterminated, farmers, wherever your Commission went, asked for jackal-proof fencing. Your Commission is of the opinion that nothing is more calculated to lead to the ultimate extermination of the jackal, than jackal-proof fencing, the erection of which should be promoted by legislation.

192. While your Commission is of the opinion that the interests of both State and farmer would best be served by making the jackal-proof fencing of farms compulsory throughout the Union, it considers the time is not yet ripe for such a step. **Where, however, in any district, the necessary majority of owners so decides, your Commission recommends that the law, which now applies to ordinary fencing only, should be amended so as to include jackal-proof fencing.**

193. **Your Commission also recommends that a law be made which will compel three or more farmers to combine for jackal-proof fencing of the boundary of their block of farms if the owners of, say, two-thirds of the area or block desire to fence the same.**

194. All owners within such an area would be called upon to pay a prorata share of the cost and maintenance of such fence, and as the owner of a boundary farm in such an area may make use of that fence for purposes more particularly his own, provision should be made for such an eventuality.

195. With regard to hunting clubs it is felt that the law is not carried out with sufficient stringency with reference to compulsory hunting, and in so far as the law allows trespass in the pursuit of a jackal, it is practically a dead letter. **Your Commission recommends that special attention be given to these points, in connection with the legislation proposed in the preceding paragraph.**

196. Since dogs are bound to play an important part in exterminating the jackal within the fenced areas, **your Commission recommends that the Department of Agriculture make the neces-**

sary enquiries as to what type or types of dog are best for hunting purposes.

197. Your Commission recognises that there are those who will consider any measure of compulsion with reference to the jackal irksome; but the benefits which will flow to both State and farmer as a result of the extermination of the jackal, far outweigh any inconvenience to which the farmer may be put. To the farmer the extermination of the jackal means large savings in the cost of herding his flocks; more and better wool; a greater freedom from stock disease and attacks by insect pests; greater protection against scab; and an increased capacity of the farm to carry stock.

198. To the State the extermination of the jackal means that the destruction of the vegetal covering of the country and soil erosion will cease, in so far as the kraaling of stock is responsible; that it will be possible to introduce the paddock system of farming, and through that bring about the restoration of the veld which at the present time is in the process of ruination.

199. **In conclusion, your Commission desires to again emphasise that the jackal is a dangerous menace to the State.**

200. **Your Commission recommends an effective campaign of extermination of the jackal and to this end suggests certain legislative measures in connection with jackal-proof fencing which are above detailed.**

XIII. PROVISION OF CHEAP FENCING MATERIAL

201. Your Commission has everywhere heard the same request for cheaper fencing material. The price of fencing material and the want of capital is undoubtedly a big obstacle in the adoption of the paddocking which your Commission has endeavoured to emphasise as one of the most important factors in reducing drought losses.

202. If the proposed jackal-proof fencing act is put into effect in the form recommended by your Commission, the State will be bound to grant loans for jackal-proof fencing. But the request for cheap fencing is not confined to the jackal-proof variety; the public desires fencing materials of all sorts at the lowest possible rates, and your Commission wishes to support them in their request.

203. It is unnecessary to reiterate the enormous advantages of fencing and paddocking. Not only will fencing improve the grazing and the stock, and assist in the extermination of the jackal,

and reduce surface erosion, but it has the additional advantage, frequently brought to the notice of your Commission, that jackal-proof fencing brings sloot formation very vividly to the notice of the farmer. The sloot which crosses a fence line is at its start, and with every increase in size, a visible and ever-increasing danger to the effectiveness of his jackal-proofing, and he is forced to give it his attention.

204. Your Commission places this request of the farming community for fencing material, on easy terms, before you for your very earnest consideration. It is a crying need in the country to-day, and the effect will be beneficial to the land which we have received from our forefathers and predecessors, and must deliver to our children and successors.

205. It is obvious from the third chapter of this report that the provision of shade and shelter by trees in the veld will tend to the improved condition of stock. This provision is usually lacking, and one of the principal reasons which your Commission has been able to deduce from the evidence is the cost of the protecting fences required during the initial stages of growth.

206. For the same reason the more extensive planting of the spineless cactus is retarded.

207. Cheaper fencing material is needed for the encouragement of these two factors in improvement of farming conditions as well as for the protection of sloots from depredations of stock while reclamation is proceeding.

208. The public requests the granting of fencing loans repayable by annual installments in twelve years at 4 per cent interest. Your Commission places the request before you in its original form and recommends that the most favourable terms possible be granted.

209. **Your Commission recommends that fencing loans be granted by the State to farmers on the best terms possible, and that under such loans farmers be permitted to put up both boundary and paddock fences, whether jackal-proof or not, and even to convert an existing stock-proof into a jackal-proof fence; and, generally, to utilize fencing for any purpose calculated to improve the drought-resisting capacity of farms.**

XIV. Development of Water Supply for Stock

210. The question of water supply has been so fully dealt with in Chapter V that it is unnecessary further to urge its importance.

211. **Your Commission recommends that the State should encourage farmers in every way possible to improve their facilities for watering stock.**

XV. The State and Soil Conservation

212. One of the most important principles, which your Commission recommends, is the adoption by the State of its responsibilities in connection with the control of soil erosion.

213. It must be clearly understood that this does not entail the filling in of dongas by the State; but merely that the State must take steps to prevent the needless waste of natural resources. Several of these necessary steps have already been described, and a general adoption of the improved methods of farming suggested will reduce soil erosion and sloot formation to a minimum.

214. For many years, however, until the desired condition of affairs is reached, soil erosion will continue to produce its evil effects, and must be dealt with.

215. One great fault in the past has been that the necessary work in connection with soil conservation was not placed in the hands of any department of State or official, and there has been no control. Your Commission considers that the time has now arrived for the termination of this state of affairs.

216. Innumerable were the requests at the meetings of your Commission for the establishment of a new Department of Reclamation for the control of soil erosion. This is one of the logical conclusions of the adoption of the responsibilities of the State in connection with soil conservation, but in view of the financial position, and the fact that the beginnings must be small, your Commission is not prepared to recommend the establishment of such a department at the present time. The immediate appointment of a competent officer to deal with these matters, however, is essential, and your Commission recommends that he be attached to the Department of Agriculture. The consensus of opinion of the farmers giving evidence in those districts, where soil erosion is serious (and they are many), is that the time has arrived for the appointment of an official who can advise them on conservation of soil. The first duty of such an officer would be educative, assisting by lecturing, by writing pamphlets, and by personal visits to farms. Included in his educative campaign would be experiments in controlling erosion on some well organised co-operative scheme

based on the past experience of the Department of Agriculture.

217. In addition to the work outlined above such a Reclamation Officer should also act as a control on the works of other Departments. That farmers are feeling the necessity for this is again evidenced in the application on behalf of several farmers to your Commission for the services of an erosion expert, in order to settle a dispute between the landowners and the Railway Board on a matter of soil erosion.

218. The question of the reduction of roads, so as to permit of paddocking, is also a matter which needs attention. The Superintendent of Roads and Local Works of the Orange Free State Province has worked out a scheme for the reduction in the number of roads, and this scheme has met with approval all over the Union. The Reclamation Officer, when appointed, will have to work in this matter in conjunction with the Provincial Authorities.

219. In a previous chapter the discussion of questions of the control of soil erosion on large catchment areas was foreshadowed. This control would naturally also be placed in the hands of the Reclamation Officer.

220. By the supply of trees, needed to stop soil erosion, at nominal rates, and reduction of railway rates on materials desired for the same purpose, the State could encourage these works.

221. The stopping of sloots and the prevention of soil erosion generally are essentially of the nature of a repair or restoration of the farm, and not an "improvement" in the specialised sense of the word. Expenditure on work of this kind is but to the farmer what "depreciation" of furniture or machinery is to the business man, and yet according to the evidence submitted by farmers to your Commission, money spent on such works may not, for income tax purposes be deducted from the gross income. Your Commission considers this absolutely incorrect even from the point of view of business, and strongly recommends that deductions of this nature be permitted. It is remarkable how strong the incentive is to carry out works and incur expenditure which may be deducted from the Income Tax returns, and this psychological phenomenon should be utilised.

222. Suggestions have been made to the effect that loans should be granted by the State to permit of sloot arresting but your Commission cannot support this suggestion as a general recommendation. The majority of the conservation works are restoration works and not improvements, they are not usually permanent, and will require a large degree of maintenance and careful watching. It

is difficult to see how the State would be justified in advancing money on such security. In exceptional cases where, by the erection of permanent masonry walls, terraced lands are reclaimed from a gutted vlei, the position may be different.

223. There is no doubt that some legislation will later become necessary to define the position between adjoining owners when anti-erosive works are to be constructed. A few problems of this nature have from time to time been submitted to your Commission, but these are not sufficient to base any recommendations on. Up to the present so little work has been done that cases of conflicting interests are as yet almost unknown. A Reclamation Officer would naturally in his experience meet with many points and, as occasion arose later, the necessary legislation could be enacted. At present none is needed.

224. **As a first step your Commission therefore recommends the immediate appointment of a Reclamation Officer who will be attached to the Department of Agriculture and be entrusted with the duties pertaining to State control of Soil Erosion as outlined in this chapter.**

XVI. PROPAGANDA AND INVESTIGATION BY THE DEPARTMENT OF AGRICULTURE

(a) **Propaganda:**

225. Your Commission found that farmers in general were unaware of existing loan facilities for fencing and other purposes, and recommends **that the Department of Agriculture take steps to bring this knowledge to the door of every farmhouse in the country.**

226. Your Commission also recommends that a sustained propaganda in favour of a natural life for small stock and against overstocking should be undertaken at once.

(b) **Investigation:**

227. With regard to overstocking and overgrazing, **your Commission recommends that the many problems connected with the grazing of stock should form the subject of thorough investigation by the Department of Agriculture, for it is only by such investigation that the best methods of management for the various types of veld can be determined.**

228. From this, it is difficult to imagine any work that is of higher importance to the permanent welfare of the Union which,

first and foremost, is a stock raising country. Many farmers told your Commission how valuable they had found the prickly pear in times of drought while from information furnished your Commission by the Grootfontein School of Agriculture, it would appear that sheep have been kept alive for over 260 days on a diet of prickly pear only.

229. Now, in many of the drier parts of the Union prickly pear is very abundant, yet is not generally used for keeping stock alive in drought times, the reason being that no cheap, simple method of dealing with the formidable spines with which the "leaves" are covered has yet been evolved. **Your Commission strongly recommends that the Government should investigate the singeing and other methods which are in use in Mexico and Texas with a view to introducing them into South Africa, for, not only is the prickly pear a source of food but also of water. The sheep mentioned above, for example, drank no water during the time of the experiment and four of them, which were subsequently put on to a diet of prickly pear and lucerne hay, have now been over a year without drinking water, the water requirements of the animals having been fully met by that contained in the prickly pear.**

11

Pastoral Development in Somalia: Herding Cooperatives as a Strategy against Desertification and Famine

Jeremy Swift

This paper describes some aspects of the recent history of the Somali nomadic pastoral economy and of the relationship in such a society between famine and desertification; the conclusions are used as a basis for suggesting broad outlines for a new type of development policy.

A pastoral ecosystem has three main components: people, animals, and land (used here as a general category to include vegetation, climate, etc.). Desertification and famine result from a breakdown in the relationships among the different components: desertification, in the broad sense of land degradation leading to a reduction in the primary vegetative yield, is the result of a major imbalance between the animals and the land; famine is the result of an imbalance between people and animals (the nomad's main means of production). This imbalance may be in absolute numbers of animals, their distribution among families, or their production or their price relative to other commodities.

In the traditional Somali nomadic pastoral ecosystem, human and animal populations were kept in a fluctuating and approximate relationship to each other and to the land by a variety of mechanisms. The major limits were set by general environmental controls such as scarce and variable rainfall, drought, scarcity of surface water, and animal diseases. The relationship was also regulated by social and economic mechanisms such as human birth regulation, redistribution of animals among pastoralists, and the spacing of grazing. As a result, major imbalances between people, animals, and land, and thus famine and desertification, were avoided much of the time. Not always, of course. Periodic drought, food shortage, feuding, and war were also major factors operating to control human and animal populations. It would be wrong, however, to see these as the only mechanisms at work in that the pastoral society's own social and economic mechanisms also played an important part in making pastoralism a viable way of life.

Recent changes, however, especially the increased marketing of Somali livestock, combined with some aspects of the formation of a modern state—first colonial, then independent—in Somalia have modified the pastoral ecosystem, and have removed some of the environmental as well as many of the social mechanisms by which this rough balance was kept. These changes have acted selectively by removing many of the old controls without replacing them, thus setting in motion major forces of economic change. As a result, concealed by a surface appearance of rapid development, the traditional pastoral society has become economically and ecologically unstable. The major famine of 1974-75 and widespread environmental destruction are signs of a state of crisis.

It is impossible, and in any case undesirable, to return to the old environmental and social mechanisms by which some order was kept in the pastoral ecosystem. Somali pastoralism must move forward. Only the government now has the authority and the means to act. In the light of this analysis I suggest that its dual task should be to change the old environmental limits and, simultaneously, to impose—or better, to create the conditions in which the pastoralists themselves can impose—more stringent social and economic controls to regulate the relationship between people, animals, and the land.

The Somali Pastoral Ecosystem

Environment

Somalia (figure 11.1) is entirely situated within the arid and semiarid zones according to Meigs' definition (1953) and has annual rainfall totals varying from a maximum of 600 to 700 mm in the south to less than 100 mm on the northern coastal plain. In

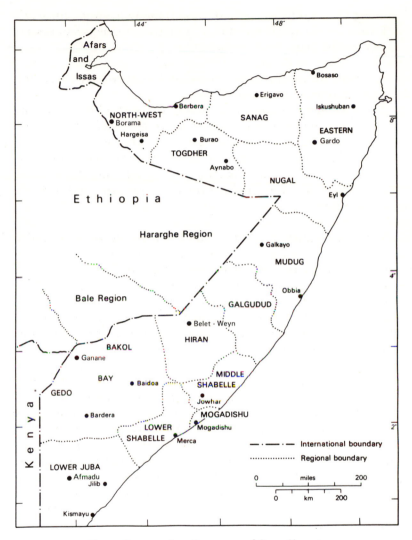

Figure 11.1. The regions and main towns of Somalia.

most of Somalia there are two rainy seasons: the main rains (*gu*) are
from April to June, depending on the southwest monsoon and the
northward movement of the intertropical front, with subsidiary
rains (*deyr*) from September to November, depending on the
northeast monsoon. The summer dry period (*xagaa*) is of variable
length and in parts of the country some rain falls throughout the

summer. The winter dry period (*jiilaal*) rarely has any rain at all.

Rainfall is extremely variable. With respect to the pastoralists it is significant that:

1. There are variations in the annual total, which affects the amount of grass that will grow
2. There are variations in the date of the onset of the *gu* (main) rains at the end of the difficult period of *jiilaal*, when a week or two of delay means increased animal mortality
3. Rainfall within a small area is also variable, which, in addition to the major movements between different rainfall zones, makes the local movement of people and animals necessary

Table 11.1 shows for selected stations the percentage probability that rain in any one year will be (a) more than half and (b) more than three-quarters of the long-term average for that station. The uncertainty of the rain, and hence of grass, is evident. This uncertainty becomes greater, the drier the area. In the driest area, with less than 100 mm mean annual rainfall, three or four years out of ten will not get even half the long-term average for that station, while in the medium rainfall areas with a long-term average annual rainfall of 200 to 400 mm, three or four years out of ten will get less than three-quarters of the long-term mean.

The onset of the *gu* rains is unreliable. Hunt (1951) shows that in northern Somalia the date of onset of widespread *gu* rains varies between 7 April and 4 May each year. Within a small area, rainfall is also unreliable. Hunt gives a short series of observations from Hargeysa at three separate stations, two of them half a mile apart in the town, with the third five miles away at the airfield (100 m altitude higher than the other two). Monthly totals differ on occasion by more than 50 mm between stations. Of the annual totals, only in one year out of three are they reasonably close; in the other two years the largest and smallest differ by 113 and 141 mm, which is a variation of 40% and 35%, respectively, for stations only five miles apart. Fantolli (1965) has published a longer series of figures from seven stations within an area of 40 km² at Jowhar. The 26-year means (1936-62) for these stations ranged from 438 mm to 534 mm, a difference of nearly 100 mm.

Meteorology as well as popular memory record the droughts of the past, especially in the north: 1911, 1914, 1918 (*Xaaraama cune*—the drought when people ate forbidden food), 1925, 1927-29

Table 11.1. Somalia, Annual Rainfall: Long-Term Means and Probabilities
at Selected Stations

Station	Period	Number of Year with Full Records	Mean Annual Rainfall (mm)	Percent Probability that Rain Each Year will Exceed:	
				75% of mean	50% of mean
Bosaso	1934–57,1964–72	20	17.3	45	55
Berbera	1906–50	43	58.0	56	67
Gardo	1939–58,1964–75	26	99.4	54	69
Galcaio	1935–58,1964–75	26	149.0	65	81
Hobio	1924–58,1964–75	29	191.8	52	90
Burao	1921–50,1967–75	34	207.3	65	94
Belet–Weyn	1926–58,1964–75	31	246.6	61	87
Ganane	1922–60	26	316.4	54	92
Bardera	1915–60,1964–72	42	374.7	62	88
Mogadishu	1910–59	44	402.7	68	93
Hargeysa	1922–50,1966–75	35	425.4	77	100
Jowhar	1921–60	38	496.9	74	97
Borama	1921–50,1967–75	25	511.3	96	100
Afmadu	1926–58	9	563.4	78	100
Baidoa	1922–60	29	584.7	72	100

Sources: Calculated from rainfall series published as follows:
Berbera, Borama, Burao, Hargeysa to 1950 (Hunt, 1951);
other stations before 1960 (Fantolli, 1965);
all stations after 1964 (Meteorological Service, Civil
Aviation Department, Ministry of Transport, Mogadishu).

(22,000 people destitute, 80% of the sheep and goats estimated to have died), 1933-34, 1938, 1943, 1950-51 (*Siigacase*—the blower of red dust), 1955, 1959, 1965, 1968-69, and the major drought of 1973-74 (Hunt, 1951; Boothman, 1975). These figures indicate drought with varying degrees of severity occurring every three to four years.

Uncertain rainfall is only one of the sources of risk to Somali pastoralists. Until the recent rinderpest vaccination campaign, this disease was a constant threat to the herds. The great rinderpest epidemic of the 1880s killed millions of animals and led to widespread starvation. Other animal diseases still continue to cause annual livestock losses.

These environmental hazards impose limits on human and animal populations and on plant production, limits which cannot be much changed by human action using only traditional science and technology. The problem is how to make an acceptably secure living within these "external" controls imposed by the environment. In a predominantly pastoral society, this means regulating the relationship between people, animals, and the land by economic, social, cultural, or political means so that major avoidable imbalances of the sort which could lead to food shortage or to environmental destruction do not occur. This regulation is accomplished by a set of economic and social mechanisms which form part of the traditional cultural ecology of the Somali pastoralists.

People and Animals

The Somali pastoral economy has been described summarily by Lewis (1961). The main unit of production is the family, grouped together with other families (rarely more than four) for part of the year. In some seasons the unmarried men and boys may form a separate herding unit which takes most of the camels away in order to exploit grazing lands far from water, leaving a few burden and milk camels, together with all the sheep and goats, with the rest of the family. Families live from the milk of their herds, from the meat of animals occasionally slaughtered, and from grain and other foods bought with money earned by the sale of animals and animal products.

The relationship between people and animals is regulated by demographic controls on the people themselves, controls which restrict the size and rate of growth of the human population, and by the demographic and economic processes which control the size of the herds and their distribution among families.

Human demography. Little good information is available on the demography of Somali pastoralists. Some demographic sample surveys have been carried out, however, three of which were among pastoralists. The results are summarized in table 11.2, which gives average household size and vital rates for urban, settled rural, and nomadic pastoral populations. There is a notable difference between the rates of natural increase of the three sectors, with nomads having the lowest rate, and urban populations the

Table 11.2. Somalia: Household Size and Vital Rates of Urban,
Rural Settled and Nomadic Pastoral Populations

Population	Average Number of People per Household	Birth Rate 0/00	Death Rate 0/00	Rate of Natural Increase 0/00
Urban	4.4	67	18	49
Rural Settled	4.7	56	34	22
Nomadic Pastoral	5.9	37	20	17

Notes: 1. Urban figures represent a weighted average of 10 towns
larger than 5,000 inhabitants, surveyed between 1967
and 1969 (one in 1962).

2. Rural settled population is a weighted average of 8 rural
farming districts surveyed between 1967 and 1969.

3. Nomad figures show a weighted average of two pilot surveys carried out in 1974.

Source: ILO/JASPA Report, 1976.

highest. This pattern, and the relevant birth and death rates in each
sector, are within the normal range found in similar conditions
elsewhere in Africa.

It should be noted that the relatively low nomadic population
growth is the result of a combination of low birth and low death
rates. This pattern is characteristic of most African pastoral
populations so far described. The mechanisms by which this is
achieved in the Somali case are not entirely clear, but it is interesting to note that, as with other African pastoralists, there appear to
be more men than women in Somali pastoral society, unlike the
ratios in other sectors of the population. Nomad sex ratios
(number of men per 100 women) in the three pilot surveys were as
follows: Bardere 110, Burao 106, Afmadow 111, while the overall
sex ratio for the 21 towns surveyed was 96. Low death rates in the
overall pastoral population in normal years may be a consequence
of the generally good levels of nutrition among nomads, with high

rates of protein consumption, and a reduction in some infectious diseases more dangerous to a sedentary community. In times of drought, of course, pastoral mortality may be high, especially among children and old people.

Several features of Somali pastoral society and culture might in fact contribute to birth control. Traditional Somali believe a woman to be fertile immediately after her menstrual period. Such factors as strict premarital chastity, high bride-wealth (leading to later marriages), polygamy, relatively easy divorce, and the temporary separation of husband and wife (common in Somali pastoral societies when husbands are traveling or helping with herds away from the wife's hut) would all contribute to slowing the birth rate.[1]

Among traditional pastoralists it was also thought desirable for intercourse to be deliberately spaced so that children were born in the best season of the year, and even for weddings to be delayed and intercourse banned during severe drought years, presumably to stop children from being born when there was not enough food for them.[2] Thus in the Somali folk story of a soothsayer whose powers of divination were tested and who, with a serpent's help, foresaw the end of a serious drought, the main sign of rejoicing is the lifting of a ban on weddings and on intercourse by married couples.[3]

People and animals: pastoral wealth distribution. Animals are individually owned, inherited, and disposed of, although there is a suggestion that (in some sense that is not very clear) camels are considered to be also the property of the clan (Lewis, 1961; Abdi, 1975). Each family lives from the produce of its animals and its fortunes depend on the size and yield of the herds. In the north of Somalia, more than half the households surveyed in 1974 owned camels, sheep, and goats, another one-third had only sheep and goats, a very small number had only camels, and a negligible number only cattle. Surveys in the Juba area in the early 1970s indicated that more than half the nomadic households owned only cattle, a relatively high proportion owned only camels, and an almost negligible number had only sheep and goats.[4]

Herds, the pastoralists' capital, are less susceptible to increase from hard work than are farmers' fields, although obviously competent and hardworking herdsmen are likely to be rewarded. Most subsistence production from the herd is in milk, which can scarcely be stored. There are thus few ways a herdsman can protect himself against the periodic risks of drought and animal disease. For this reason, in the traditional pastoral ecosystem, economic differentiation was slowed by the high degree of environmental risk and by the lack of ways in which a pastoral surplus could be invested either in risk-free assets (Lewis, 1968) or in productive capital to increase the yield from the herds themselves. Economic institutions seem in any case to have ensured a partial redistribution of herds from rich to poor at times of special need.

Two such mechanisms were of particular importance in traditional Somali pastoral society. A family, losing its animals, was supported by kinsmen and friends in the following way (Hashi, 1969). The victims were at once given some lactating goats and camels, as well as some fat rams for immediate slaughter, to provide meat to help them recover from the shock of the disaster. A few days later a general meeting was called at which all families were represented. This meeting established the facts of the loss, especially whether the loss was due to negligence of the owner or to causes beyond his control A small group of men was then chosen to decide on the contribution of each family, a contribution which was usually composed of a two-year-old sheep, a three-year-old goat, or up to four head from every 100 sheep and goats. If the original loss was the fault of the owner, the minimum quantity was given together with a strong warning. If he was not at fault, half or all his losses were made good.

In less dramatic cases, men without enough animals to be independent could work as herdsmen for richer pastoralists, for urban livestock merchants, or for settled farmers who owned animals but were unable to migrate with them. Standard herding contracts with payment in cash or kind existed in different parts of the country.[5] Although such arrangements opened the way for the exploitation of labor and the creation of a permanent class of hired herdsmen, they do not seem to have had this result in the traditional Somali pastoral economy. The fact that hired herdsmen were often paid in animals seems to indicate that such arrangements were principally a means for impoverished families to acquire the necessary pastoral resources to become again independent production units. In practice also they seemed to have worked this way.

Animals and Land

The mechanisms just described—in particular, voluntary and involuntary human demographic controls and the various mechanisms for redistributing animals according to need within the pastoral society—regulated the relationship between people and animals. Another set of mechanisms regulated the relationships between animals and the land. Like the first set, they were partly deliberate and partly involuntary.

Somali pastoralists keep camels, cattle, sheep, and goats. These animals eat different parts of the vegetation, thus making good use of the diversity of plants. In terms of ecological efficiency they come somewhere between the grazing pattern of a full fauna of wild animals, which simultaneously exploit a large number of different ecological niches, and the pattern of a large modern ranch, where one species of domestic animal eats only one part of the vegetation. For example, camels and goats, which can exist on thorn bush alone and which graze in a dispersed pattern—with the camels grazing far from water—have a less heavy impact on the land than do the sheep and cattle which eat mainly grass, are gregarious, and stay close to water.

In the traditional setting, there seem to have *once* been formal mechanisms, amounting to *de jure* control of grazing. For example, Cerulli (1959) records that traditional northern chiefs had the power to declare a particular pasture closed (*xirmo*). For a given period of time it was forbidden to graze livestock there and special penalties enforced this decision. The declared purpose of this was to let the grass grow. The Sultan had a similar power to declare forested hills (*xayran*) off limits, and to prevent tree cutting for a given period of time. Wood belonged collectively to the clan in whose normal territory it was found, while hillsides with gum arabic, frankincense, or myrrh were individually owned and the harvest rights sold by the owner.

However, no such mechanism for grazing control is recorded by more recent research. Individual clans occupied particular areas of pasture because of their ownership of wells. Access to water, population numbers, effective occupancy and, in the final resort, military superiority alone guaranteed access to pasture. As a result, water availability and military solidarity between kinsmen and allies were the most important regulators of grazing pressure. During the dry season, animals had to be watered regularly (camels every eight days or so, the other species much more often), and so could not be far from water. This led to *de facto* control of pasture use, resulting from a fairly well-distributed pattern of grazing far from the home wells in the wet season, with the animals gradually falling on the ungrazed areas closer to the wells as the dry season advanced.

It should be noted that the mechanisms regulating the relationship between animals and land in the traditional Somali pastoral ecosystem were relatively weak and ineffectual. Somali pastoralism is characterized by relatively low productivity and by fluctuating and unpredictable resources. In general, the system has not proved to be an adequate economic base for the creation of powerful political organizations with the kind of central organization that can impose stringent land-use discipline. But the uncertainty of the environment itself has been a powerful means of control. Periodic droughts and disease epizootics decimated the herds regularly and generally kept them in a long-term dynamic equilibrium with the vegetation's ability to regenerate itself.

Development of Commercial Pastoralism

Although the camel boys sometimes live for several months on camel milk alone, Somali pastoralists cannot, or do not choose to, live by their herds alone. They eat millet, sorghum, and rice, and need clothes, tea, sugar, and other goods they do not produce themselves. They sell animals and animal products or commodities they have hunted or gathered in order to buy these items.

Early nineteenth century trade in northern Somalia seems to have concentrated on clarified butter or ghee (*subag*) and on wild products (ivory, myrrh, ostrich feathers, gum arabic) collected by the nomads, and was apparently not concerned with substantial numbers of live animals. This pattern of exports was changed and trade was considerably expanded by the English occupation and development of Aden situated on the opposite coast across the Gulf of Aden. By 1869, for example, Berbera supplied all the livestock consumed by the British garrison and the local inhabitants. Live animals quickly became a major item in trade.[6]

The development of commercial pastoralism in the twentieth century in Somalia is illustrated in table 11.3. These figures should be treated with more than normal caution, since apart from the usual statistical uncertainties, such as unregistered exports in small boats from the northern coast, there are two major sources of bias: (a) the figures for northern exports include an unknown quantity of animals from Ethiopia, variously estimated at 10 to 40% of Berbera exports, and (b) the figures for southern Somalia do not include a considerable number of cattle which are exported illegally through neighboring countries, especially Kenya. This flow is estimated at between 35,000 and 60,000 head annually (UN/ECA and UN/FAO, 1972).[7]

Table 11.3. Somalia: Development of Commercial Pastoralism; Exports of Animals and Animal Products in Selected Years

	1919	1927	1936	1950	1959	1963	1972	1975
Northern Somalia:								
Sheep and Goats (head)	61,483	127,544	118,462	119,610	440,054			
Cattle (head)	1,675	2,274	2,172	1,311	5,629			
Camels (head)		18	650	174	3,613			
Ghee (metric tons)	23	251	150	2				
Hides and Skins (thousand)	1,239	1,755	1,301	1,548				
All Somalia:								
Sheep and Goats (head)				120,875	454,602	828,726	1,635,779	1,536,361
Cattle (head)				2,651	14,185	39,951	81,328	39,883
Camels (head)				174	3,613	15,302	21,954	34,223
Ghee (metric tons)				385	255		26	
Hides and Skins (metric tons)					3,644		4,912	

Sources: Northern Somalia: Annual Colonial Reports, London, HMSO.
Southern Somalia: 1950 and 1959 - Rapport du Gouvernement Italien
à l'Assemblée Générale des Nations Unies sur l'Administration de
Tutelle de la Somalie. Rome: Ministère des Affaires Etrangères,
Instituto Poligrafico dello Stato, 1956 and 1959.
Somalia 1963, 1972, 1975: Somali Democratic Republic, Central Sta-
tistical Office and Livestock Development Agency.

Nevertheless, these incomplete figures give an indicative picture of the commercial development of traditional Somali pastoralism. Despite large annual fluctuations in exports—probably due, in part, to variations in rainfall and pasture—there has been a sustained growth in the export of pastoral products since the first decade of this century. The most important element has been a great increase in the export of live animals. Sheep and goat exports grew slowly through the 1920s and 1930s, and suddenly accelerated in the mid-1950s (1953 in the south and 1954 in the north). The main reason was rapidly increasing purchasing power of their main customer, Saudi Arabia, as well as of other oil-producing Arab states.

The pattern of commercialization of traditional nomadic pastoral production in Somalia can thus be summarized as follows. The pastoralists' cash needs seem to have been met in the early nineteenth century by marketing wild products they had either hunted or collected, such as ivory, myrrh, ostrich feathers, and gum arabic, together with ghee and some animals. After the British occupation of Aden and the northern Somali coast, large animals and hides and skins, which went largely to Europe and the United States, began to dominate the trade. For a time trade in hides and skins was most important. The live animal trade, however, grew rapidly after the mid-1950s, as a result of the aforementioned increased purchasing power in the Arabian peninsula and the Gulf states, displacing hides and skins as the main pastoral product. Growth in the export of camels and of goats was particularly fast, while the export of ghee has declined sharply. These changes in the composition of pastoral exports are shown in table 11.4. In 1972, the great majority of live animals went to Saudi Arabia (86% by value) and to the People's Democratic Republic of Yemen (8%).

The Consequences of Increased Commercial Pastoralism

The growth of livestock marketing in Somalia has had far-reaching consequences for the pastoral economy, for the pastoral producers, and for the pastoral environment. This increase, I contend, does not reflect a sustained increase in livestock production. Rather it has been accompanied by the beginnings of major changes in the pastoral society, particularly in increased economic stratification, increased private appropriation of resources, and a breakdown—without adequate modern replacement—of some of the communal institutions and cultural mechanisms which had

Table 11.4. Somalia: Pastoral Exports' Share of Total Exports
(percent by value)

	1950		1973
	North	South	All Somalia
Live Animals	27	0.9	58
Hides and Skins	67	27	3.8
Ghee	0.04	5.7	0.2
Total Exports	100	100	100

Sources: 1950 North Somalia – <u>Colonial Office Report</u>. London: HMSO
1950 South Somalia – <u>Rapport du Gouvernement Italien à l'Assemblée Générale des Nations Unies sur L'Adminis-tration de Tutelle de la Somalie</u>. Rome: Ministère des Affaires Etrangères.
1973 Central Statistical Department, Mogadishu. <u>Sta-tistical Abstracts 1973</u>.

previously regulated economic life and ecological organization. As a result, pastoral producers are increasingly vulnerable to destitution and famine. In addition, there has been an uncontrolled use of water and grazing, leading to environmental destruction.

Within the pastoral economy it is important to distinguish between two different sets of productive activities: subsistence production in which animals are managed to provide food direct-ly, mainly in the form of milk and also to some degree meat; and market production in which animals are sold or bartered to provide food grains, cloth, and other commodities. These two types of production are partly complementary to each other: young male animals are not normally needed for food production and, in Somalia, have limited value as transport animals, so they can be sold, while animals too old to breed can also be sold without affecting food production. They are also partly in contradiction: animals sold could be eaten instead; more important, young animals and people compete for milk—if people take less milk the survival rate of the young animals improves because they have more milk, and market production benefits; if people take more milk, their own nutrition improves but young animals suffer, mortality increases, and there are fewer to be marketed. This potential contradiction is highlighted by the fundamental difference between the objective of a traditional pastoral economy, which is to support a large number of people with an acceptable safety margin against disaster, and the objective of a modern high-productivity animal production operation such as a ranch, which is to market the maximum quantity of milk or meat, and which requires as few people as possible competing with the calves for milk, or consuming part of the lamb or calf surplus. The transition between one type of pastoralism and another, even ignoring the social and political implications of such a profound economic transformation, obviously means a major change of management strategy and economic behavior. In a difficult and hazardous environment such as the Somali rangelands, a transition of this kind is fraught with evident dangers. Mere emphasis on marketed offtake at the expense of subsistence production is in itself unlikely to accomplish satisfactorily the necessary transformation.

In the event that livestock marketing by Somali pastoralists was a response to increases in external demand, it was not accompanied by sustained capital investment or by planned changes in the organization of production. The main investments were in veterinary care and in the provision of water. For example, there was rapid development of water supplies in the rangelands, especially after the mid-1950s, both by government and by private individuals, but there were few other changes in production. In particular, there was almost no successful attempt to introduce modern forms of grazing control: grazing reserves were tried by the British colonial administration, and failed; subsequent attempts at forming grazing associations, and at creating grazing reserves in a few areas of the north, were overtaken by the 1974 drought, although they had met with some success. There were no general or technical education programs for nomadic pastoralists, nor were there extension programs or provisions of credit. In all, there were no improvements in any aspect of the livestock production economy other than water and disease control and, as a result, there was no increase in the productivity of labor and land. In these circumstances, the boom in livestock marketing, starting in the early twentieth century and more especially since the mid-1950s, has not been the result of development in any real sense. Rather, it has been a combination of a semimodern marketing operation imposed on a largely traditional production system. It meant a windfall gain to traders and to the government through export taxes, but this was accomplished at the expense of an increasingly destructive exploitation both of the pastoral food economy and of the pastoral environment.

However, livestock marketing set into motion social, economic, and ecological changes in the pastoral economy. The growth of the marketing of livestock and livestock products gave rise to a new social class—livestock merchants—an urban group with strong kinship and contractual links with both the nomadic producers and with the emerging administrative and later governmental elite (Lewis, 1962). The livestock trading activities of this new class had important consequences for the pastoral economy. In particular, increased marketing activities began to break down economic solidarity between kinsmen and began to replace a variety of nonmarket transactions by market transactions. For example, the traditional mechanism by which individual herdsmen and families in trouble were helped through a redistribution of the corporate wealth became less widespread as animals were withdrawn from circulation within the traditional pastoral economy in favor of sale outside it. The growth in camel marketing, a particularly striking feature of the 1950s, made progressively less possible their earlier role as the joint wealth of the lineage group. As cash incomes grew, successful individuals, especially urban merchants, needed less group solidarity and were less ready to contribute to it. Meanwhile, less successful individuals were more rarely able to call upon collective solidarity and, consequently, were less able to escape their poverty. The beginnings of a process of economic and social differentiation appear in the formation of this wealthy urban livestock trading class and in the progressive breakdown of the former collective obligations, as well as of the social and economic organization of the traditional pastoral society.[8]

The growth of commercial pastoralism and the increasing monetarization of the pastoral economy also led to increased private appropriations of what had previously been communal resources. There has been some fencing of pasture and also a rapid increase in private cultivation of fodder for use by animals held for export in the pens at Berbera and on board ship. The most widespread consequence, however, has been a rapid development of additional water supplies in the rangelands since the mid-1950s. Some of these were government-built boreholes and surface water tanks, but the majority have been private, cemented rain water tanks (*barkads*), equipped with petrol pumps, built by livestock merchants, rich pastoralists, and Somali sailors who returned from abroad with money to spend. Drums of water are also now taken to remote pastures by lorry, and sold to pastoralists for their animals with the price fluctuating widely with demand. This increase in water availability has had important ecological consequences. For example, animals can stay longer at the start of the dry season in previously waterless areas. The fact that water is pumped removes the limit to herd size previously imposed by the need to draw water by hand. As a result, grazing pressure has increased. Since many of these new water sources are open only to those who can pay for the water rather than being the collective property of the kinsmen who dug them, they have spearheaded the intrusion of new pastoral groups (nomads with no traditional pasture rights, traders with market herds, and newly settled farmers with cattle) into grazing areas which were previously under the *de facto* control of groups with traditional water rights in the area. These factors have led to a further breakdown of traditional grazing discipline.[9]

Private water tanks have also tended to reinforce both the position of the livestock traders, who thus became monopolistic suppliers of water in some areas, and the power of those pastoralists who could afford to pay for water. In practice, the wealthy pastoralists ensured their own access to grazing in the vicinity of the water, while those who could not pay for water were excluded. The new water supplies also led to some partial settlement by pastoralists who started growing sorghum and millet. Cattle at such settlements fed for part of the year on crop residue, but for the rest of the year grazed close to the settlements. This led to considerable ecological damage from an almost permanent grazing pressure which was unrelieved by seasonal nomadic movement away from the area. Since this settlement by pastoralists has usually taken place in the best watered areas (close to the rivers in southern Somalia and in the wadis or *togs* and flood plains in the north), some of the best pasture, often that used only in the dry season, has been removed from grazing.

Planning the Future

I have identified as some of the main elements of the Somali pastoral crisis: the development of an increased marketed offtake from the herds, not accompanied by increasing productivity of livestock production, which has resulted in putting domestic food production and further export earnings at risk; ecological disorder with respect to land use, with a disappearance of the old *de facto* and *de jure* grazing controls without adequate replacement; disappearance of the traditional social security system by which herdsmen were protected against the risks of an unreliable environment; and the beginning of economic stratification. The combination of these factors has contributed in recent years to a major famine and widespread desertification. Although, with the return of adequate rainfall, pastoral production will no doubt recover somewhat and desertification will for a time be less evident, there seems little doubt that the Somali pastoral economy and its environment are in an increasingly precarious position, vulnerable to drought and unable to contribute, as they should, to the development of the Somali economy as a whole.

Future planning for the pastoral sector has to start from these present conditions. In Somalia the pastoral sector is the most important part of the economy since over two-thirds of the national population are nomadic pastoralists, and since exports of animals and animal products bring in about 70% of foreign exchange earnings, a proportion which is likely to grow as a result of the crisis now faced by their other main export commodity, bananas.

The Somali government is a socialist government, committed to transforming a dependent colonial economy into a modern self-reliant socialist one. This political choice implies a planned transformation of production techniques and organization, production relations, class relations, and the exercise of political power. In a predominantly pastoral economy this is a considerable challenge. It means finding new technologies appropriate to the task, to the limited resources available, and to the technical competence of the users. It also means finding new forms of social and economic organization within which this technology can best be used.

One aspect of the new strategy has been the settlement of 150,000 pastoralists in farms and fisheries in the south of the country. Initially, this was undertaken as an emergency drought relief measure. The experience gained during this experiment will aid in the future planning of agricultural development schemes, The simple population arithmetic of Somalia, however, indicates the size of the task. Table 11.5 shows estimates of population growth in the different economic sectors in the next five years.

Table 11.5. Somalia: Sectoral Population Growth 1977-81,
and its Possible Absorption

Sector	Assumed Proportion of Total 1976 Population (percent)	Assumed 1976 Population ('000)	Assumed Annual Growth Rate (percent)	1977-81 Natural Sectoral Population Increase ('000)	1977-81 Sectoral Population Increase After Proposed Inter-sectoral Transfers ('000)
Pastoral Nomadic	65	2,120	1.7	186	0
Settled Agriculture	15	489	2.2	56	216
Fisheries	1	33	2.2	4	30
Urban Services and Industry	19	620	4.9	168	168
Total	100	3,262	2.4	414	414

Source: ILO/JASPA Report, 1976.

At present population growth rates, the Somali population will increase by 400,000 people in the next five years. Of these, 186,000 will be in the pastoral sector, despite the fact that it has a lower growth rate than other sectors. The ILO/JASPA team concluded that since industrial employment was unlikely to increase at a much greater rate than the urban population during this period, all the rural population increase would have to be absorbed in the countryside. A viable strategy would seem to mean transferring the necessary number of people (186,000 over five years) from pastoralism to agriculture and fisheries, so that the pastoral sector does not continue to grow in absolute size. With the recent experience of the first agricultural and fisheries settlement schemes, it appears realistic to plan for the bulk of the increment of the pastoral population to be transferred into agriculture and the rest into fisheries. The figures involved are shown in the last column of table 11.5.

The creation in the next five years of adequately capitalized agricultural and fisheries schemes, with sufficient trained staff, for about 250,000 people—the majority of whom will be former nomadic pastoralists—is a major task for the Somali government. Valuable experience is being gained in the implementation of such schemes and some experience from other countries is relevant. However, the main planning task, for which there is little experience from other countries to guide the government, is to develop pastoralism itself, since it will remain the main occupation of most people in the country for some time, as well as the main source of foreign exchange. If the nomadic population can be held constant during the next five years, and even progressively reduced thereafter, planning becomes easier. Since the size and economic importance of Somalia's nomadic pastoral population is unique in Africa, solutions are called for that are specific to Somalia. New comprehensive policies for pastoral development have to be defined, policies which take account of the complexity of the pastoral ecosystem as well as the legitimate demands of the pastoralists. New organizational forms have to be invented by which these policies can be put into effect. Some recent discussion on this latter problem has centered on the idea of pastoral cooperatives (see, for example, ILO/JASPA, 1976).

An important aim of such policies and of pastoral cooperatives would be to reestablish more effective controls over the relationship between people, animals, and land, and thus to tackle the problems identified in this chapter as the sources of the Somali pastoral crisis. Other possible development policies and organizational structures such as private or state ranches, partial investment in some aspects of production such as water or disease control, or the creation of grazing reserves unrelated to other aspects of livestock production and marketing do not appear to be able to meet all the diverse requirements that have been discussed. Since there is no experience of pastoral cooperatives in Africa, it is worth considering some of the tasks such cooperatives might undertake.

Economic Objectives of Pastoral Cooperatives

The main economic objective is to increase the productivity of pastoralism, so as to create a pastoral economy less subject to the vagaries of nature within an organizational framework that accomplishes at least three linked tasks: (a) to guarantee the subsistence of all herdsmen in bad years, (b) to find a balance between production for herdsmen's own consumption and production for market and export, and (c) to reverse the incipient processes of economic differentiation within the pastoral society as well as between the countryside and the towns.

The main technical means to increase productivity in a pastoral economy are well known. They consist principally of techniques already in use in Somalia such as water development, including water spreading and water harvesting, disease control, and fodder production and livestock fattening. Some of these techniques have contributed to the pastoral crisis in Somalia, but this is not so much the fault of the technique itself as the lack of an adequate institutional framework by which technical changes are controlled and related to other parts of the pastoral ecosystem, particularly to resource use. Pastoral cooperatives would provide such a framework.

One of the main tasks of the cooperatives would be to provide security to herdsmen in bad years through the sale of food at fixed prices. This would be accompanied by a program of buying animals at fixed prices which would enable the herdsmen to switch more easily from pastoral to vegetable foods and thereby to destock the range rapidly.[10] The cooperative might operate an insurance fund against individual loss of animals, and could provide basic banking and credit facilities. The cooperatives might be in charge of all livestock marketing (in conjunction with existing government agencies such as the Livestock Development

Agency) and retail sales of basic commodities to pastoralists. Profits from these operations—especially livestock marketing— would be an important source of cooperative finance.

Cooperatives might also grow food crops and fodder, fatten animals for sale or slaughter, and process and sell animal products (milk, ghee, hides, and skins). The cooperative center would be a base for emergency food storage and for experiments in storing meat and milk. Some complementarities with existing agricultural and fishing cooperatives, for example in the exchange of produce, might be profitable. Within individual cooperatives there would be scope to organize labor during the slack season for self-help schemes in construction and in land and water conservation. From the early days of the cooperatives it would probably be necessary to constitute special *ad hoc* groups of paid workers for agriculture, construction, and specialized tasks such as machine operation and maintenance, artisan work, carpentry, and mini-pastoral industries such as hides, skins, or milk processing. These jobs should be rotated between families as far as possible.

A major policy decision needs to be made on the extent to which communal herds would replace private herds. It would probably be premature to envisage early transition from one system to the other. The organization of this should grow slowly with experience, and as the pastoralists gain confidence. It should be remembered that enforced communalization of animals was the obstacle on which the first phase of Mongolian pastoral cooperatives failed (Humphrey, 1977). It may be advantageous to keep private ownership of herds for the foreseeable future, but to tax animals in excess of a certain figure per family (100 sheep and goats, and 50 camels, for example), a modest but sharply rising amount. Alternatively, it might be possible to reward pastoralists for limiting herds to specific sizes. The cooperatives' own herds should be built up slowly over the years, allowing the formation of specialist herding brigades.

Ecological Objectives of Pastoral Cooperatives

The main ecological objectives of cooperatives would be to halt environmental deterioration, reverse the trend until optimum range and water use is achieved, and maintain this over the long term; in other words, to ensure that increasing productivity of the pastoral environment is achieved without degrading the environment. The essential task is to provide a framework within which the herdsmen's use of the range and water, especially at the site of new technical developments such as wells, is better controlled and directed. Improvement is impossible in the present near-anarchy use of range and water. An organizational framework must be provided so that changes and investments can be made by the state and by the pastoralists themselves for improving land, water, and range—changes not subject to the normal problems of collective property abuse which arise through unrestricted access leading to overuse or neglect.

A particular group of people has to be made responsible for a particular area of land, in which they and the state can make improvements that will not be appropriated by other pastoralists. Pastoral cooperatives would give management and surveillance responsibility for the rangelands to specific users, thus replacing competitive individual decision making with collective responsibility.

The most appropriate size for such an area must be determined by trial and error. In Mongolia the pastoral cooperative is identical in territory to the smallest administrative division of the state and contains 400 to 500 families (Humphrey, 1977). As a general rule, in an area of very uncertain rain like Somalia, the larger the management unit, the greater the chance of good rain somewhere within its boundaries. It might be best to start with large territorial units encompassing the broad boundaries of present nomadic movements of identifiable groups and gradually to subdivide, as management experience grows and the confidence of the pastoralists increases.

In fact, since water distribution is the key to grazing, the manipulation of water supply would be as important as territorial boundaries. Varying combinations of permanent and temporary water resources can be used to distribute grazing in a desired pattern, thus recreating and redefining in modern terms the traditional grazing discipline. The purpose would be to distribute animals as widely as possible in the wet season (*gu* to *deyr*), leaving permanent water and adequate pasture within reach for the *jiilaal*, dry season. The dry season water source has to be closed or otherwise forbidden to animals until *jiilaal* starts. The total number of animals in the cooperative would be calculated each year on the carrying capacity of dry season range within reach of permanent water. For the rest of the year, temporary water sources—natural surface pools, or *barkads*—would be used for watering when necessary. Each cooperative would have several of these permanent/temporary water systems. In order to gain control of water distribution, the state may have to acquire control of all water sources by nationalization or other means.

The most difficult first step would be to delimit the cooperative boundaries and to assign pastoral families to them. This stage especially would necessitate fundamental consultation with the pastoral families, not just between technical experts and government officials. After cooperative boundaries were established, cooperative members would be required to stay within them, although it might be desirable in some circumstances to allow for reciprocal grazing rights between cooperatives. Members would inform local police authorities of any transgression of cooperative boundaries by members of other cooperatives.

Since camels range farther than other animals and eat trees and bushes seldom consumed by the others, it may be desirable not to limit camels to cooperative territory in the wet season from *gu* to the end of *deyr*, but to let them follow their traditional wide-ranging grazing patterns. During *jiilaal*, however, they would be restricted to cooperative territory.

Within the cooperative territory, the cooperative administration would undertake range management and pasture improvement, including such technical measures as bunding; would set aside seasonal grazing reserves, forest reserves, and perhaps forest plantations; and would possibly institute wildlife conservation measures. The cooperative administration would have the job of determining year by year the approximate grazing pressure acceptable in the dry season. The mechanism by which cooperative herds would be reduced in years of low carrying capacity would be important. The cooperative would probably have to buy the excess animals at fixed prices, with the purchases spread proportionately among members.

Social Objectives of Pastoral Cooperatives

The principal social objectives of pastoral cooperatives would be to ensure adequate participation by pastoralists in the life of the nation and to bring them the benefits of a modern state. The essential task is to provide an organizational structure that will allow the state to act coherently in the pastoral sector in order to implement social policies, to provide education and health services to nomads, and to make possible the participation of nomads in national political and economic processes, particularly in the framing of the development plans which so radically affect them.

The cooperatives should offer members positive economic and social advantages as the reward for obeying grazing rules, since there can be no question of forcing Somalia's pastoralists into a strict new organization against their will. Pastoralists have already shown themselves ready to abide by the rules in the grazing reserves in northern Somalia. With an attractive package of advantages there is every reason to think that the people will join and abide by the rules. At the start it might be an idea to encourage people with few animals to join the cooperatives by making them loans of animals—perhaps using the same rules as the traditional animal loan system described above—on the condition that they abide by cooperative rules.

Each cooperative would have, probably near the dry-season wells or suitable agricultural land, an administrative and service center with a school, human and animal health services, and a cultural center. Cooperative members would receive training here during *jiilaal*, but would nomadize with their families during the rest of the year. In this regard, radios might be used successfully to back up the basic education and health services. The cooperative center itself might be somewhat mobile in the first few years until the most suitable site is agreed upon.

One of the cooperative's main functions, and its primary attraction to pastoralists, would be the provision of basic education and health services. The difficulties of providing such services to nomadic people have never been fully overcome but the successful Iranian experience with tent schools and nomadic front-line health workers merits study in this respect. It would probably be desirable to combine fixed installations at the cooperative center with nomadic school teachers and health workers, radio educational programs, and possibly radio links between field health workers and cooperative medical centers. The fact that in a pastoral economy, herdsmen are employed for much of the year principally in supervising their animals rather than in active labor, means that there is considerable dead time that could be used for radio or cassette-tape educational programs.

The cooperative would be the main vehicle of state administration and intervention in the pastoral sector. Thus, apart from its economic functions, it would be the means of giving nomadic pastoralists political education, consulting with them on policies to be followed, and providing feedback to the central political organization—in other words, involving the pastoralists as far as possible in decision making at every level. Since there is little experience of pastoral development anywhere in the world, there is a special need for this feedback which will enable mistakes to be spotted early enough for correction.

In conclusion, I have suggested that the crisis in Somali pastoralism has resulted in part from the breakdown of such traditional institutions as were available to regulate the relationship between people, animals, and the land, and by the manifest inadequacy of such institutions, particularly in a rapidly modernizing nation. Traditional pastoral political, economic, and ecological organization has been weakened, leaving a vacuum in the countryside that has not yet been filled effectively by the state. Pastoral cooperatives would be an appropriate form of organization to take on some of the functions of traditional pastoral society, such as the regulation of grazing, security against loss of animals, regulation of conflicts over land use, and making investments in the land. Pastoral cooperatives would also be an appropriate vehicle for taking the state to the nomads and for making their views and wishes known to the state, a channel of two-way communication that is needed particularly during a phase of rapid transformation of the sort now starting in Somalia.

General Note

The material for this paper was gathered during two visits to Somalia, the first as a member of a UN/IDEP (African Institute for Economic Development and Planning) mission, the second as part of the International Labour Office/Jobs and Skills Programme for Africa (JASPA) team headed by Dr. S. Nigam. Some of the discussion in this paper draws on material entitled *Economic Transformation in a Socialist Framework: An Employment and Basic Needs-Oriented Development Strategy for Somalia* (ILO/ JASPA, Addis Ababa, October 1976, in press). This document is referred to as "ILO/JASPA report" in references for this chapter. I am grateful to colleagues of these missions, and especially to Somali colleagues and friends for information, discussion, and criticism. The views expressed here are my own and do not in any way commit the United Nations agencies concerned, nor do they necessarily reflect Somali government views. They are offered as a contribution to the healthy debate now under way on pastoral planning in Somalia.

Notes

1. In addition there appear to have been, and still appear to be at least in part, specific birth regulation devices. In traditional Somali pastoral society, sexual intercourse was banned during the period of one and one-half to two years when the wife was breast feeding the new baby; during this period the husband slept in the *ardaa* just outside the hut. This ban used to be universal and is still practiced by some pastoralists, although it is not clear to what extent. Such a ban is an effective birth control measure and has the added advantage of being homeostatic in its action. Since it lasts only so long as the mother is lactating, an early child death removes the ban. During a period of economic plenty child survival can be expected to be high and thus birth spacing is increased; during a period of high infant mortality due, for example, to drought, birth spacing is reduced and the birth rate rises to compensate (Somali Academy of Culture, author's interviews).

2. There is a story of a woman who only gave birth during the dry *jiilaal* season when there was little milk from the animals. She composed the following poem to her children, complaining of this:

> Isna ma waantoobo wacdaraale aabbahaa,
> Anna ma waantoobo waaga roob ma dhalo.

> Your troublesome father doesn't learn from experience,
> And I never learn from experience: I never give birth
> in the rainy season.

3. "Tell him that the wives who were banished from the companionship of their husbands during the hard season will make huts as large as stone houses. . . . Tell him that they will spread sleeping-mats in a snug recess of the hut. Tell him that the husbands who had no thought for love in the hard season will now enter the huts. . . . Tell him that as the wives pass to and fro close to their husbands, the love which had grown old will become young again. Tell him that in their revelry and play they will conceive sons of blessing, bright as thunderbolts. Tell him that the unmarried men will wed, in a befitting way, the girls of their choice." (From "The Soothsayer who was put to the Test." I am grateful to Dr. B. W.

Andrzejewski, who translated this story, for bringing it to my attention.)

4. The slender data available on size of pastoral family holdings in Somalia are summarized in the ILO/JASPA report. These figures are based on a very small number of surveys taken under difficult conditions and too much credence should not be put in them. They do, however, seem to indicate plausible orders of magnitude, in light of pastoral economies elsewhere. In the north, the great number of family sheep and goat holdings range between 50 and 150; there are very few below 25, and few above 150, although there is a minor peak of holdings above 300. It seems reasonable to conclude that around 50 sheep and goats are the minimum necessary for family subsistence and that 150 to 200 represents some upper limit, probably a limit on available labor. This limit may be exceeded only by the few families rich enough to hire labor.

The surveys show that few camel herds exceed 50, except for a small number of large herd owners with several hundred. These figures bear out popular hearsay which claims that in Somalia 10 to 20 camels is a sufficient number for a family, that a man is rich with 100, and that a few men own over 1,000. The surveys also showed a negligible number of nomads owning cattle in north Somalia, while in the Juba region of southern Somalia there is a marked peak in cattle holdings of 21 to 50 cattle per household.

5. For example, in the Burao and Hargeysa Region, a hired herdsman received from the owner a food ration (cereals, tea, sugar), two garments a year, and the milk from the herd, in addition to 300 Somali shillings (So. Sh.) per six months for looking after 50 camels, or 100 sheep and goats, or 50 cattle. In the south, a hired herdsman worked by annual contract and received one she-camel or one two-year-old cow, whichever was relevant, plus the milk from the herd, and clothes twice a year. Herdsmen who stayed in the village and did not take the herd into the mountains were usually paid in cash at a monthly rate of 1.50 to 2 So. Sh. per cow, or 0.50 to 0.60 So. Sh. per sheep or goat (International Labor Organization, 1972).

6. In one year (1875-76) Aden imported 63,000 sheep and goats, and 1,100 cattle from Berbera. In return the Somalis received grain, sugar, dates, iron, beads, and especially cotton. Few camels were recorded as trade, although some were sent to India, presumably for army use. Trade from southern Somalia developed more slowly. In the 1870s Mogadishu exported ivory and other products from the interior, grain and gum from the coastal regions, and hides and skins, but few live animals (Pankhurst, 1965a, 1965b).

7. An additional note to table 11.3: Sheep and goat exports from northern Somalia grew more than 15% per year between 1950 and 1959, and exports from southern Somalia during the same period grew much faster, although from a small base. Sheep and goat exports from Somalia as a whole grew by over 10% per year between 1959 and 1972, before exports were affected by the severe drought of the mid-1970s.

Exports of cattle were low before the recent Arabian boom in the mid-1950s. Since then they have grown rapidly, though somewhat erratically, with large annual differences in animal exports. Recorded live cattle exports grew by 14% annually between 1959 and 1972; if recorded slaughterings were to be included the rate would be higher. Although these figures are certainly not an accurate record of herd exploitation given the large, fluctuating, and uncontrolled exports on the hoof through neighboring countries, there has clearly been a considerable increase in cattle marketing.

Camels were not exported in significant numbers, nor were they otherwise marketed on a large scale, until the mid-1950s when there was a rapid rise in exports. Recorded camel exports from northern Somalia, the main camel region, increased from 174 in 1950 to over 3,600 in 1959. Camel exports from Somalia as a whole increased by nearly 15% per year between 1959 and 1972. In addition, large numbers were slaughtered for meat. Unlike cattle, sheep, and goats, camel exports do not seem to be as adversely affected by weather. While exports of all other species fell in the years of drought 1972-75, camel exports rose, and camel exports do not fluctuate as much as other species from year to year.

8. This network of collective obligations between kinsmen and contractual partners, which had hitherto been a means of protection against economic and ecological uncertainties and against the political rivalries of competing clans, came in fact to dominate, in a deformed way, the development of economic and political institutions in the new Somali state. People used their kinship and contractual links with others in powerful positions. Politics and economic development were quickly dominated by "tribalism" of this sort. The government which came to power on 21 October 1969 made it one of its main aims to eliminate this.

9. It is relevant to note that traditional northern Somali pastoral movements had earlier been adversely affected by the British colonial authorities, when in 1954 they had recognized Ethiopian sovereignty over the pastures of the Haud, pastures essential to large numbers of Somali pastoralists.

10. The cost of such an operation would have to be carried by

cooperative marketing levies and other funds, with central government help in bad drought years. The cost of this would almost certainly be considerably less, even in cash terms, than the huge cost of relief operations such as that in the 1974-75 drought. In addition, the social benefits would be incalculable.

References

Abdi, Muse, 1975. In *Proceedings of the SIDAM/IDEP Seminar on Somalia in Transition*. SIDAM, Mogadishu, Somalia, November (in press).

Boothman, I. M., 1975. An historical survey of the incidence of drought in northern Somalia. In *Abaar, the Somali Drought*. I. M. Lewis, ed. London, International Africa Institute.

Cerulli, Enrico, 1959. Il diritto consuetudinario della Somalia settentrionale (Migiurtini). In *Somalia, Scritti Vari Editi ed Inediti*, Vol. 2. Rome: Instituto Poligrafico dello Stato per l'Amministrazione Fiduciaria Italiana della Somalia.

Fantolli, A., 1965. Contributo alla climatologia della Somalia: Riassunto dei risultati e tabelle miteorologiche e pluvio metriche. Roma: Ministero degli Affari Esteri.

Hashi, Mohamed Abdi, 1969. Country paper on Somalia. ILO Technical Meeting on the problems of nomadism in the Sahelian region of Africa. Niamey, 9-20 September 1969. Mimeo, 11 pp.

Humphrey, Caroline, 1977. Pastoral nomadism in Mongolia; the role of herdsmen's cooperatives in the national economy. *Development and Change*. The Hague, Nederlands (in press).

Hunt, J. A., 1951. *A General Survey of the Somaliland Protectorate*, 1944-1950. Hargeisa, Somalia: Chief Secretary. 203 pp.

ILO, 1972. *Report to the Government of the Somali Democratic Republic on the Integrated Development of the Nomadic Zones*. Geneva: ILO/OTA/Somalia/R.6. 67 pp.

ILO/JASPA, 1976. *Economic Transformation in a Socialist Framework: An Employment and Basic Needs-Oriented Development Strategy for Somalia*. Ethiopia: Addis Ababa, October (in press).

Lewis, I. M., 1961. *A Pastoral Democracy: A Study of Pastoralism and Politics among the Northern Somali of the Horn of Africa*. London: Oxford U.P. (for the International Africa Institute).

Lewis, I. M., 1962. Lineage continuity and modern commerce in northern Somaliland, pp. 365-385. In P. Bohannan and G. Dalton, eds., *Markets in Africa.* Northwestern University Press, Evanston. 762 pp.

Lewis, I. M., 1968. Land Tenure Conditions. In *FAO Agriculture and Water Surveys, Somalia.* Final Report, Vol. 6, *Social and Economic Aspects of Development.* FAO, Rome.

Lewis, I. M., ed., 1975. *Abaar, the Somali Drought.* International Africa Institute, London.

Meigs, P., 1953. World distribution of arid and semiarid homoclimates. In *Reviews of Research on Arid Zone Hydrology.* UNESCO, Paris.

Pankhurst, R., 1965a. The trade of the Gulf of Aden ports of Africa in the nineteenth and early twentieth centuries. *Journal of Ethiopian Studies,* Addis Ababa, 3(1): 36-81.

Pankhurst, R., 1965b. The trade of southern and western Ethiopia and the Indian Ocean ports in the nineteenth and early twentieth centuries. *Journal of Ethiopian Studies,* Addis Ababa, 3(2): 37-74.

UN/ECA and UN/FAO, 1972. United Nations Economic Commission for Africa (UN/ECA) and Food and Agriculture Organisation (FAO) of the United Nations. *Prospects for Production, Marketing and Trade in Livestock and Livestock Products in Eastern Africa to 1985.* Vol. 2, *Country Studies.* Addis Ababa: October.

Whitely, W. H., 1964. *A Selection of African Prose. 1. Traditional Oral Texts.* Oxford: Clarendon Press, 200 pp.

12
Climate and Weather Modification in and around Arid Lands in Africa

Michael H. Glantz

This paper presents some of the climate and weather modification schemes that have been discussed seriously since 1900, especially during times of prolonged meteorological and agricultural droughts and dry spells in the arid and semiarid regions in Africa (Glantz, 1977). These schemes were seriously proposed by their respective authors as "scientific" solutions either to combat the desertification process, to minimize the impact of periodic extended droughts or to reclaim land lost to desertification, or all of these. Some of the proposed modifications, such as the construction of tree belts and cloud seeding, have become operational in several African countries and elsewhere in the world. In addition, the asphalt island experiment was almost carried out in Venezuela in the late 1960s, but was scrapped because of internal Venezuelan politics (Kelley, 1968).

There are many who believe that technological advancements or technology transfer will resolve many of the environmental prob-

This chapter is an updated, expanded version of a paper by Michael Glantz and William Parton, "Weather and Climate Modification and the Future of the Sahara," in M. H. Glantz, ed., *Politics of a Natural Disaster: Case of the Sahel Drought* (New York: Praeger Publishers, 1976), pp. 303-324. It was presented at the International Geographical Union's Working Group on Desertification In and Around Arid Land Symposium (Ashkhabad, U.S.S.R.), July 20, 1976.

lems faced by inhabitants and governments in Africa. There are others, however, who believe that technological solutions must be scientifically assessed before their application for their social as well as physical implications. Meanwhile, there exists much information which, if properly used, could improve the human situation on the African continent.

The recent prolonged drought in the Sahel has renewed interest in resolving the perennial problems that confront arid and semi-arid regions in Africa and elsewhere. The Sahel, in fact, could supply good examples of problems in the arid lands of developing states. The situation along the southern fringe of the Sahara will be described briefly.

The Sahara

The Sahara, currently the world's largest desert, extends from the Atlantic Ocean across to the Red Sea. In Africa it touches about 13 countries comprising a total population in excess of 150 million.

The Sahelian region extends from west to east and straddles the geographic border between the southern Sahara and the sudan zones; it is part of the savannah belt which receives a long-term average annual rainfall of 200 to 600 mm. It is a marginal climatic zone and as such is subject to wide variations of precipitation in both time and space. The Sahel encompasses six former French colonies—Senegal, Mauritania, Mali, Niger, Upper Volta, and Chad—states that, along with adjacent states such as Nigeria, have been suffering from the effects of the six years of drought that ended in 1974.

During the drought, precipitation in the Sahel was considerably less than its long-term averages. The sharp reduction in rainfall had a major impact on the Sahelian nomads, who lost the livestock on which they depended for food and transportation, which in turn forced them to settle in makeshift camps close to population centers. Sedentary farmers unable to produce adequate, harvestable crops became dependent on the urban centers for whatever welfare their governments or international donors might be able to supply. Those in urban centers, affected by the influx of refugees from other parts of the region, had to look to the international community for food relief (Winstanley, 1975). Although the drought has since ended, new food emergencies in the Sahel are still disrupting long range development plans for the region, with relatively scarce resources being diverted to stopgap relief measures (*Rocky Mountain News*, 1976).

It is important to keep in mind that the regions surrounding the Sahara, especially the Sahel, are potentially productive but that poor land management practices such as indiscriminate well drilling, overgrazing, and reduction of fallow time have reduced the ability of this "marginal" land to withstand the impact of periodic extended dry spells and droughts.

Most of the weather and climate modification proposals for the Sahara and the Sahel have been suggested in an attempt to make the land arable again and to compensate for the tendency of the inhabitants to destroy their land through poor land management. With respect to the former objective, it has been suggested, for example, that there exists a large reservoir of water under the Sahara Desert:

> Men have only recently become aware of the vast dimensions of the subterranean reservoir below the Sahara. . . .
> good soils are also available; recent investigations have shown that extensive tracts of desert land have arable (and once cultivated) soils overlain by a thin cover of sand (Ambroggi, 1966).

With respect to the latter objective, it is generally accepted that the agricultural and animal husbandry practices of the region's inhabitants have had a negative impact on the fertility of the land. Nicholas Wade recently summarized this view:

> The primary cause of the desertification is man, and the desert in the Sahel is not so much a natural expansion of the Sahara but is being formed *in situ* under the impact of human activity (Wade, 1974; see also Glantz, 1975).

Although there have been three droughts in West Africa since 1900, a major difference in the setting of the first major drought and the recent one involves technological capabilities. Awareness of the growing capability of the scientific and engineering communities to modify weather and climate, and the potential lack of political restraints on such potential modification experiments, has come at a time when the global climate appears to be undergoing certain unexplainable variations. To some, these variations mark the beginning of a transition to a new climate norm; to others, they are just fluctuations within the current climate regime. Regardless of which view proves correct, the effects of these variations have been felt around the world, especially in 1972, the year of anomalies: excessive droughts in sub-Saharan Africa, India, the Soviet Union, Central America, Brazil, and East Africa; and excessive rains in the Philippines, New Zealand, and parts of Kenya and Australia. In addition, major grain exporters like Canada, the United States, and Australia also were affected by adverse climatic fluctuations in 1974, reducing the world's available grain reserves at a time when populations were still increasing at alarming rates.

The impact of these events coupled with revelations of military employment of weather modification during the recent wars in Indochina (Shapley, 1974) has drawn attention to the possible social and economic consequences of planned weather and climate modification. While some members of the scientific community take a hopeful view of modification techniques, others are extremely skeptical of their value. Figures 12.1 and 12.2 illustrate several engineering schemes that have been proposed to modify or to control the climate. The possible impact of some non-African schemes has been discussed by Kellogg and Schneider (1974).

When discussing modification schemes, it is important to distinguish between weather and climate modification. Weather modification is generally taken to mean that the impact will be of short duration. Cloud-seeding operations designed to produce a change in precipitation serve as one example. The impact of climate modification, on the other hand, extends over a longer period of time (a month, a season, or more). Melting the Arctic icecap by spreading black particles on the ice would be an example of climate modification.

Types of weather and climate modification schemes can be further divided into two categories based on the intent of the modification: advertent (planned) and inadvertent (unplanned). The former is intentionally undertaken to achieve a predetermined objective, such as cloud seeding to suppress hail. The latter refers to activities that are designed to achieve a specific objective—not necessarily the modification of the weather or the climate—but which unintentionally trigger side effects that modify weather or climate. An example of inadvertent weather modification would be the effect of urban areas on precipitation. Huff and Changnon (1975) have shown that "measurable enhancement of precipita-

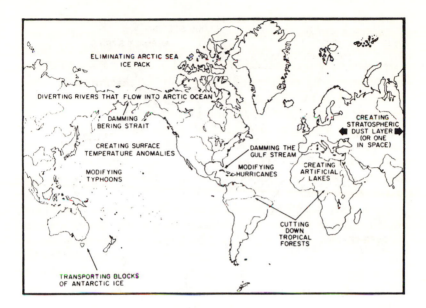

Figure 12.1. Engineering schemes that could be or have been proposed to modify or control the climate. Reprinted with permission of W. W. Kellogg and S. H. Schneider, National Center for Atmospheric Research (Boulder, Colorado).

tion has existed for several years in and/or downwind of six large urban centers in the humid portions of the United States."

For convenience, weather and climate modification proposals relating to arid and semiarid lands in Africa can be divided into the three following, not necessarily mutually exclusive, categories based on the techniques used for the purpose of advertent modification: (a) vegetation modification, (b) atmospheric circulation modification, and (c) precipitation modification.

Vegetation Modification

Weather and climate modification schemes relating to a change in vegetative cover are based on the assumption that there exist feedback interactions between the biosphere and the atmosphere. A 1950 UN/FAO report succinctly stated this view:

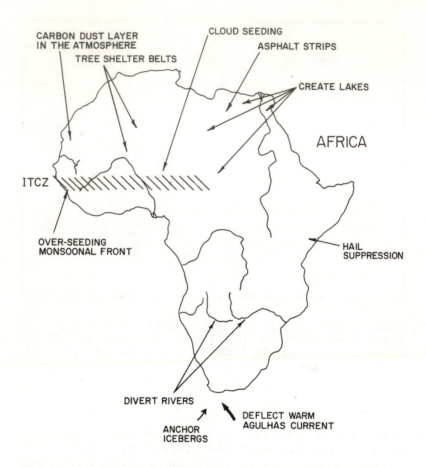

Figure 12.2. Engineering schemes that have been proposed to modify the climate or the weather in and around arid lands in Africa.

The influence of forests on climate, however, is on less pure ground, and leads to considerable controversy. Nevertheless, extensive areas of woodlands such as the equatorial forests in Africa appear to have a definite and positive effect on climate, and the destruction of such woodlands would undoubtedly cause climate disturbances that might eventually make the Continent as a whole uninhabitable (quoted in Grant, 1973).

In the 1930s, Stebbing, a British forestry professor, wrote several articles based on this point of view. He suggested that deforestation, coupled with overuse of the land through poor land management practices, had led to desiccation of the soil. This, in turn, had led to a slow but sure decrease in annual rainfall in the region, said Stebbing, who observed the deterioration of the ecosystem to be as follows:

woodlands → savannah → sahel → desert

According to Stebbing (1938), desiccation of the soil

> is held to be primarily due to the over-utilization of the vegetative covering of the soil, under which productivity is reduced, water supplies decrease in the springs, streams, rivers and wells, the water table sinks in the soil strata and the rainfall decreases.

In other writings he referred to a linkage between the vegetative cover and rainfall, but did not elaborate. Owen Lattimore (1938), in an essay on geographical factors and Mongol history, referred to this linkage in a similarly general way:

> Both the cultivation of marginal areas and the overgrazing of stock in true steppe areas can ruin the soil, create deserts and "change" the climate.

Stebbing (1935), concerned about the shift southward of the desert sands at an estimated rate of 1 km per year, suggested the construction of a forest belt across West Africa. The purpose of the belt was to stop the sands from being blown by the northeasterly winds onto arable land to the south, and to break up the hot, dry winds that desiccate the soils over which they pass. In addition, the belt would encourage moisture retention in the soil and, for some reason unexplained by Stebbing, increase rainfall. This shelter belt would have the same purpose as the one that was to have been developed in the 1930s in the United States.

With the current drought in West Africa, there has been renewed interest in a "green barrier" (tree belt) across the northern and southern fringes of the Sahara Desert and an extensive 20-year tree-planting program was begun in 1975 in Algeria. When finished, the planned barrier will be 950 mi long, 10 mi wide, and will extend from the Moroccan to the Tunisian border. It is designed to stop the northward advance of the Sahara and to reclaim 70,000 mi^2 of steppe for farming and stock raising (ALESCO/UNEP, 1975; *The New York Times,* 16 February 1975). Such schemes are based on the assumption that they will stop the destructive impact of the hot, dry winds, reduce erosion, increase rainfall, reduce runoff and evapotranspiration, and improve water intake (Le Houérou, 1976).

The value of such schemes, however, has been questioned by Le Houérou (1976), who noted that scientists, climatologists, geophysicists, and ecologists, among others, agree that "in arid zones reforestation cannot have any significant effect on the regional climate." He concluded, also, that the planting of solid strips of trees across desert edges without subsequent management organization is purely mythical and doomed to costly failure; an observation he based on economic, financial, management, and technical considerations.

Recent scientific research by Schnell (1974), Charney (1975), and Otterman (1974, 1977) on the question of feedback mechanisms between vegetation and the atmosphere suggests ways in which the feedback processes might operate. For example, Russell Schnell (1974) has stated:

> Evidence has recently . . . shown that (very small) organic particles (i.e., biogenic nuclei) produced during the decay of vegetable matter may be important in providing nuclei in clouds upon which ice may form and from which precipitation may develop.

Schnell's study suggests that the destruction, by natural and human causes, of the vegetative cover will lead to inadvertent climate modification because of a decrease in the availability of biogenic nuclei.

Another recent attempt to analyze the linkages between the vegetative cover and its impact on rainfall and climate modification is a climate simulation experiment performed with a mathematical model by Jule Charney et al. (1975). About this experiment they have written:

> Two integrations of a global general circulation model, differing only in the prescribed surface albedo (reflectivity) in the Sahara, show that an increase in albedo resulting from a decrease in plant cover causes a decrease in

rainfall. Thus, any tendency for plant cover to decrease would be reinforced by a decrease in rainfall, and could initiate or perpetuate a drought.

If the hypotheses of Schnell, Charney, Otterman, and others concerning inadvertent climate modification are proven valid (they all operate in the same direction), Stebbing's idea about the linkage between deforestation and decreased rainfall would be scientifically strengthened.

Atmospheric Circulation Modification

This group of modification schemes is designed to alter the physical characteristics of the atmospheric boundary layer, the kilometer or so of the lower atmosphere that is greatly affected by conditions at the earth's surface. Three such schemes are the asphalt island concept, carbon black dust in the atmosphere, and the creation of large inland seas in Africa.

Asphalt Island Concept

There is usually water vapor in the atmosphere over arid or semiarid lands, but a mechanism is lacking to bring it out as rainfall until it reaches distant mountain ranges. James Black (1970) suggested that significant additional rainfall can be produced in regions by spraying a large area near a body of water with a thick coating of asphalt. The surface albedo (reflectivity) of the coating would be much lower than that of the surrounding sand, soil, vegetation, or water. The temperature difference between the coated surface and its uncoated surroundings would lead to vertical convection of air over the blackened strip. The heated asphalt surface would heat the air column above it, and the heated air would rise. The rising air motion would promote cloud formation and, ultimately, precipitation—if sufficient moisture were present in the atmosphere, there would be no need for the experiment to be done near a body of water (Black and Tarmy, 1963). This technique, as shown in figure 12.3, is designed to produce an effect similar to that caused by mountain ranges, but its magnitude would be less. "This effect is particularly striking on the Mediterranean coast of North Africa where a general rainfall of about 76 mm per year increases to 500 to 800 mm in

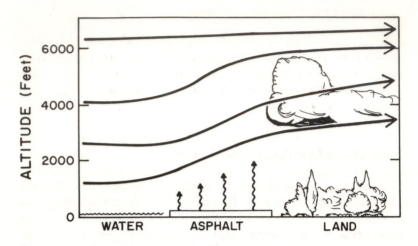

Figure 12.3. Schematic of model used to match thermal mountains to real ones.

the neighborhood of mountains" (Black and Popkin, 1967). To reinforce his proposition, Black (1970) cited as potential (heat) sources of precipitation large cities, "cloud streets" (cloud trails associated with tropical islands), large fires, and large industrial areas.

It was suggested (Black and Tarmy, 1963) that this proposed weather or climate modification process could produce "2 or 3 acres of arable land per acre of asphalt" and that "the investment cost of producing the heavy rainfall region downwind from a weather modification coating [was] estimated to be about $75 per acre (as of 1967)" as opposed to irrigation project costs in Egypt and Australia from $300 to $600 per acre (Black and Popkin, 1967).

It should be pointed out that, although other studies reinforce the heat source cloud formation proposition (Purdom, 1973; Malkus, 1963), the idea of weather or climate modification attributed to the construction of asphalt islands has been based on intuition, theoretical considerations, and analogy, and has yet to be operationally verified (Gray et al., 1974). It also appears that to date no country has been willing to allow tens or hundreds of square miles of its land to be asphalt-coated for the purpose of increasing rainfall downwind of the asphalt, regardless of how barren the land in question.

Carbon Dust

Although not specifically recommended for application in Africa, the carbon dust proposal is similar to Black's asphalt island concept. The major difference is that the artificial heat source would be placed directly into the atmosphere rather than on the earth's surface. W. Gray and colleagues (1974) have proposed that "beneficial mesoscale (100-200 km) weather modification may be possible in the coming decade or two by solar absorption of carbon dust" in order to enhance rainfall along tropical and subtropical coastlines or to enhance cumulonimbus (rain cloud) formation over areas in need of precipitation. This would be accomplished by placing carbon dust in the atmosphere to absorb solar radiation, which would then heat the surrounding air. The carbon black, acting as an artificial heat source, would increase evaporation—if released over a water surface, for example—from the water surface and promote extra cumulus convection over the land area. Gray and colleagues assume that the increased evaporation and convection would cause an increase in rainfall over the coastal area. They suggest (1974) that the carbon dust heat source is more efficient than the asphalt island source because: (a) the dust is mixed in and moves with the air the experimenters are seeking to heat; (b) when applied over the ocean, nearly all the solar energy absorption by the carbon dust is extra energy gained; and (c) the area the dust can impact will be much greater (100 km² for asphalt as opposed to 10,000 to 100,000 km² for carbon dust). Gray (1976) more recently noted that 1 to 2 million kg of carbon black would be required for weather modification on this scale.

A preliminary environmental impact statement on this use of carbon dust has recently been reinforced by a somewhat favorable (but to this author alarming) environmental assessment by Gray (1976) on the possible long-term effect of the accumulation of this dust in the environment. These effects would be multiplied severalfold because the experiment has to be reinitiated each time the heat source is to be created. The asphalt strips, however, would have a one-time construction, but would require some degree of periodic maintenance. The carbon black technique of weather modification has not yet been tested.

Finally, it should be noted that placing carbon dust in the

atmosphere could lead to the opposite of the desired effect. It could lead to atmospheric warming at high enough elevations to suppress cloud formation. That is, it could lead to increased atmospheric stability, if the dark-colored dust absorbed a significant fraction of the solar energy that would otherwise have penetrated to the earth's surface, since surface heating drives the convection instability that causes much of the initial rainfall.

Creation of African Seas

Authors throughout the twentieth century have suggested the creation of large inland seas in Africa. Such seas were supposed to supply the atmosphere with moisture which would then be returned as rainfall in the vicinity of the seas. Such a cycle would increase—at least in theory—the agricultural potential of these otherwise arid and semiarid regions. It was suggested that the seas could be created by damming or diverting the flow of major rivers into existing inland (drainage) basins, the locations of which are shown in figure 12.4.

In a survey article on potential climate modification schemes, Fletcher (1971) referred to the creation of two inland seas—the Congo and the Chad Seas—which would cover about 10% of the African continent. With respect to the proposed Chad Sea, it was suggested by Sergel (quoted in Rusin and Flit, 1962) that

> Most of the waters of this sea would be used to irrigate the Sahara, while the surplus would be channelled into the Mediterranean through an artificially created second Nile.

This scheme is shown in figure 12.5.

Other similar, but relatively more popular, schemes have been suggested concerning the development of large inland bodies of water in and around the Sahara and the Kalahari deserts. These schemes (to be discussed below) have received the periodic interest of some members of the scientific and political communities, an interest which has been regenerated by the occasional recurrence of extended droughts in West and South Africa, respectively.

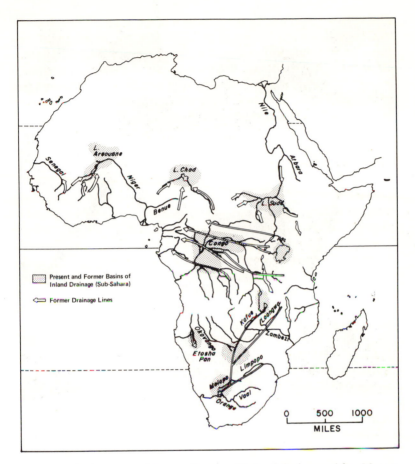

Figure 12.4. Present and former basins of inland drainage. After Mount-joy and Embleton, 1967. Reprinted with permission of Praeger Publishers.

Lake Sahara

Since about a quarter of the whole Sahara Desert area lies below sea level, the construction of a canal some fifty miles long through the higher land of the north African coast would immediately create a Sahara Sea equal in size

Figure 12.5. Sergel's plan for the irrigation of the Sahara. From Rusin and Flit, 1962.

to about half the extent of the Mediterranean. . . . All the arid regions now surrounding the desert and those parts of the Sahara which are now above the level of the ocean would be rendered as fertile as Europe. . . . Millions of human beings could then support themselves in comfort, who now lead a miserable existence on the verge of starvation (Thompson, 1912).

This statement was made by Professor Etchegoyen, a distinguished French scientist who, about 1910, attempted to foster interest in altering the climate of the African continent to enhance "its value as a place of colonization for Europeans." The sea, he suggested, would lead to an increase in precipitation. Differential

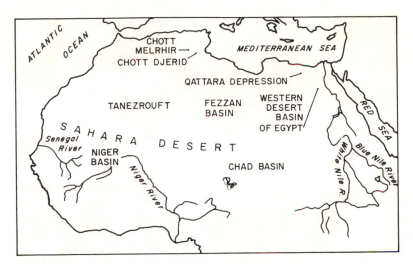

Figure 12.6. Depressions in the Sahara.

heating of the body of water and the surrounding land would create convection, which in turn would lead to increased precipitation, in part because increased water vapor due to evaporation would exist in the area above the lake and around its edges. One author referred to such a scheme as the Loch Ness monster of North Africa (Le Houérou, 1973). Figure 12.6 shows the location of various depressions in and around the Sahara.

A recent U.N. Environment Program report (1975) alluded to the fact that ecological studies were apparently being undertaken for the Qattara Depression Project in Egypt, one of the goals of which would be as follows: A canal will be cut across the isthmus separating the depression from the Mediterranean Sea, and a large-scale hydroelectric power station will be built at the precipice of the depression.

James MacDonald (1962) challenged some of the basic assumptions behind such projects as Lake Sahara. In his article, "The Evaporation-Precipitation Fallacy," he questioned the supposition that

> the creation of open water areas from which water would evaporate to augment the atmosphere's stock of vapor . . . will lead to increased local precipitations. In the driest period of a severe drought in any part of the world, huge

masses of water are still drifting invisibly overhead, the principal cause of drought almost invariably being lack of dynamic processes capable of producing ascending motions that cause adiabatic cooling and, hence, cloud-formation as the indispensable first steps in getting any of the water down to the earth's surface.

In 1974 Rapp and Warshaw undertook a numerical model simulation to determine the impact of a relatively small modification in surface boundary conditions (modification of the earth's surface). The Mintz-Arakawa model that they used divided the atmosphere into 3,240 grid points. Lake Sahara covered an area encompassing only six of the model's grid points. Globally, this modification is small, but on a regional basis it is quite large. While noting the shortcomings of the model on which the simulation was based, they concluded that there was a very low probability (<0.05) that there would be a change in the rainfall around the lake but that there would be increased precipitation in a mountainous region about 900 km from the lake. It should be emphasized, however, that there have been many problems associated with the model used by Rapp and Warshaw. Therefore, their results should be treated accordingly (Gates, 1975). As they noted, their findings were consistent with those presented above by MacDonald, asserting that although there may be an increase in atmospheric moisture caused by the presence of a large body of water in an arid region, there is still a need for a precipitation mechanism to release the water into the atmosphere. The moisture in the atmosphere may have to be transported 100 to 1,000 km before reaching a condition favorable for the occurrence of precipitation. On this point Peixoto and Ali Kettani (1973) concluded that because of moisture transport ". . . the theory of the formation of precipitation from evaporation in a given place cannot be accepted."

It has also been suggested that there may exist favorable side effects with the creation of Lake Sahara. Peixoto and Ali Kettani (1973) pointed out that

> if Chott Melrhir [a large depression in Algeria and Tunisia] were flooded entirely by water from the Mediterranean, the discharge of the Great Eastern Erg [in Algeria] through evaporation would be greatly reduced, thus limiting the loss of usable underground water.

It is possible that combinations of these weather and climate modification techniques could be made to further enhance their individually stated goals. For example, one might combine the Lake Sahara concept with the asphalt island theory in an attempt to further increase precipitation in the ways that each of these techniques alone is (theoretically) expected to produce. Before any of these atmospheric circulation modification techniques becomes operational, however, an assessment of their implications should be systematically undertaken. Such questions as, "Where will the displaced descending air go?" or, "What impact will it have on the strength of the monsoonal front from which West Africa, including the Sahel, gets its annual precipitation?" must be answered at least in theory.

Schwarz Scheme (The Kalahari)

E. H. L. Schwarz (1920), as early as 1918, suggested a scheme designed to modify the atmospheric circulation in the arid regions of what were then territories under the political dominance of Great Britain and South Africa. He suggested that, diverting the flow of existing river systems away from the ocean and "back" into ancient inland drainage basins would reconstitute large lakes in the Kalahari, thereby reestablishing an ancient "internal circulatory system." This system was based on Schwarz's belief that, as had been the possible case in the past, one could

> obtain sufficient water from the Chobe and Okavango Rivers to fill the Makarikari. This water, saved from the sea, [would] be evaporated, and [would] fall as rain during the next year.

This existence of the reconstituted inland lakes, coupled with the hypothesized concomitant increase in rainfall, would open up otherwise useless land for irrigation schemes, as shown in figure 12.7. The later followers of Schwarz stressed the scheme's irrigation potential more than its precipitation modification aspect.

Various aspects of the proposed scheme, as well as the assumptions on which it was based, have been both directly and indirectly questioned. MacDonald's comments on the evaporation-precipitation fallacy, for example, apply. Doubt has been voiced

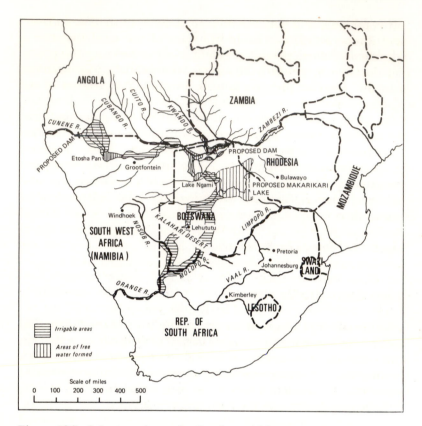

Figure 12.7. Schwarz scheme for Southern Africa.

that any appreciable increase in rainfall would be generated in the area to be reclaimed by the reconstitution of the Kalahari lakes. One reviewer (Anonymous, 1925) noted by analogy to other existing river-fed basins that "if more water is turned into the Kalahari, we should expect the outgoing air to contain more water vapor" and that "the whole question depends on the extent to which water evaporated from the proposed lakes will be again precipitated within the basin."

One of the major reasons prompting Schwarz to make his proposal stemmed from the belief in the early part of the 1900s that rainfall in the Union of South Africa was decreasing annually. Although the controversy continues today, the evidence appears to reinforce the view that South Africa is not suffering from a meteorologically-induced trend toward desiccation (Tyson, Dyer, and Mametse, 1975). The justification of the need for such a scheme, therefore, becomes questionable. Furthermore, the political situation in southern Africa has been radically altered in the past few years in such a way as to preclude the creation of the Kalahari lakes.

In 1920 Schwarz had also noted that he had to exclude the Zambesi River from his project for political reasons because "no meddling with Victoria Falls would be allowed." Today, however, the political problems are considerably more complex. The region encompassed by the project would involve coordination and agreement among at least the following four independent or potentially independent states—the Republic of South Africa, Botswana, Angola, and South West Africa (Namibia). Such agreement and coordination would be highly unlikely, given the contemporary politico-military setting in southern Africa.

Other South African Schemes

Other schemes suggested for the modification of the climate and weather in South Africa include a variety of methods for deflecting the warm Agulhas current toward the western coast of South Africa in an attempt to increase precipitation along the arid and semiarid western coast of that country.

Another scheme, one apparently taken seriously in the 1950s, called for transporting Antarctic icebergs to the western coastal waters of South Africa. It was hypothesized that in addition to serving as a source of drinking water, the "very humid warm air circulating there would be chilled sufficiently to cause increased precipitation over the coastal belt and maybe further inland" (Taljaard, 1976). A recent suggestion to tow icebergs from the Antarctic to the Arabian coast has apparently received more serious consideration by government officials in Saudi Arabia (*Christian Science Monitor*, 1975).

Precipitation Modification

Perhaps the most popular form of weather modification is precipitation modification by cloud seeding for the purpose of either (a) augmenting precipitation or (b) suppressing hail. These techniques have been widely discussed in scientific journals and meetings as well as in the popular press. Cloud seeding operations for sub-Saharan Africa have received some attention in past decades but interest in them has increased as a result of the recent Sahelian drought. Such operations will be discussed below. Hail suppression operations, occurring in regions with more favorable precipitation, will not be discussed in this paper (Alusa, 1974; Atlas, 1976).

Precipitation Augmentation

After more than 30 years of experimentation with cloud seeding to augment precipitation, apparently the best that a U.S. National Academy of Sciences report (1973) on this weather modification practice could conclude was that

> on the basis of statistical analysis of well-designed field experiments ice-nuclei seeding can sometimes lead to more precipitation, can sometimes lead to less precipitation, and at other times the nuclei have no effect, depending on the meteorological conditions.

It has generally been accepted that there is available moisture in continental clouds as well as in maritime clouds, especially in coastal regions. For those interested in precipitation augmentation, the basic problem becomes one of increasing the clouds' efficiency with respect to processing water vapor in the atmosphere. With both types of clouds, the fact remains that only a relatively small fraction of the water vapor in an air mass falls out as rain. Cloud seeding is used to increase precipitation from cloud systems that "were already producing natural rain or were about to produce rain."

There has been growing interest in the potential uses of precipitation augmentation in general, and specifically for rain-deficient regions of the world such as the Sahel. At a recent meeting of the World Meteorological Organization (WMO) Working Group on Cloud Physics and Weather Modification of the Committee on Atmospheric Science (1974), a "Draft Project Proposal for the Artificial Augmentation of Precipitation in the Sahelian Zone" was prepared. It proposed that the project take six to seven years and be divided into three phases: feasibility study, preliminary experiments, and, if these are encouraging, an operational phase.

The project would evaluate the feasibility of precipitation augmentation in the Sahel, and the results would determine its feasibility as a weather modification scheme for other arid regions. Although it would take at least two years, if not several more, to evaluate the use of precipitation augmentation for the Sahelian states, commercial cloud-seeding activities have already been carried out in the region.

For example, Weather Science, Inc., a private American firm, was contracted by the government of Niger to undertake emergency cloud-seeding operations in September and October 1973. The seeding design used in "Project Rain—1973" was based on research conducted in the United States by several groups interested in this weather modification technique. Ray Booker, the firm's president, reported that only storms in the extreme southern section of Niger (Niamey, Maradi, and Zinder) that had some potential for producing rain were seeded with the hope of augmenting such rainfall (see figure 12.8).

The process was described by Booker (1973) as follows:

> The target clouds were selected for seeding when, in the judgment of the pilot, they met the general criteria. . . . In practice, this meant selecting actively growing (cloud) towers which were believed to be destined to rain in some amount.

Booker's conclusions about the cloud-seeding operations in Niger were optimistic, though he noted that "no evaluation of the effectiveness of seeding [was] possible on such a brief project." It was not clear in the report, however, on what he based his optimism.

Cloud seeding had also been employed in Senegal as early as 1954. Drought-like conditions in 1966 prompted a renewed interest by Senegal in precipitation augmentation. The results of these subsequent efforts, however, have been unpromising (Seck, 1974).

Cloud-seeding operations have been carried out in such countries as the U.S.S.R., the United States, Brazil, Canada, France, Israel, Japan, Australia, and Zaire with varying degrees of declared success. Figure 12.9 indicates nations that have experimented with and/or conducted cloud-seeding operations (Changnon, 1975; Colorado International Corporation, 1975). Several important and timely questions on the feasibility and effectiveness

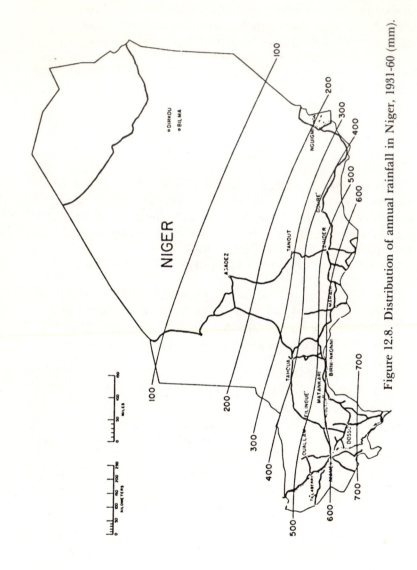

Figure 12.8. Distribution of annual rainfall in Niger, 1931-60 (mm).

of this form of weather modification have been raised. For example, a National Academy of Sciences report (1973) has recently queried:

- What is the increase in rainfall that can be expected from cloud seeding?
- What effect does cloud seeding have on the internal dynamics of clouds?
- What are the extra-area effects of cloud seeding?

These questions point to the need for additional scientific research, such as that research which was once proposed by the WMO for the Sahel.

Recently, an idea for water transfer from the water-rich tropical parts of Africa to the water-poor regions, like the Sahel in West Africa, was brought to light (Falkenmark and Lindh, 1974). This plan would involve

> large-scale redistribution of prevailing rainfall by displacement of the tropical front . . . resulting in more rain in the interior of African than at present. It would . . . be of interest to suppress part of the copious rains in Cameroon, Nigeria, and other coastal states in North Africa (sic) to keep more moisture in the monsoon for condensation further inland (Bergeron, 1970).

This could be accomplished, Bergeron suggested, by overseeding clouds in areas of high rainfall, along the Gulf of Guinea coast, for example, so that the rainfall would be suppressed, only to be precipitated out of the atmosphere further inland over the relatively arid zones.

The degree of accuracy of his assumption that the overseeding of clouds in high-rainfall areas can cause rainfall to be suppressed and precipitated at a later time has yet to be determined. In fact, Bergeron (1970) recognized this when he suggested that

> in order to lay a quantitative foundation for such investigations and experiments, it would be advisable to perform a study of the water-vapour balance over Africa . . . and to use tritium as a tracer for following the exchange of air in this region.

Bergeron also suggested that widespread irrigation across the savannah and steppe of sub-Saharan Africa, using major African

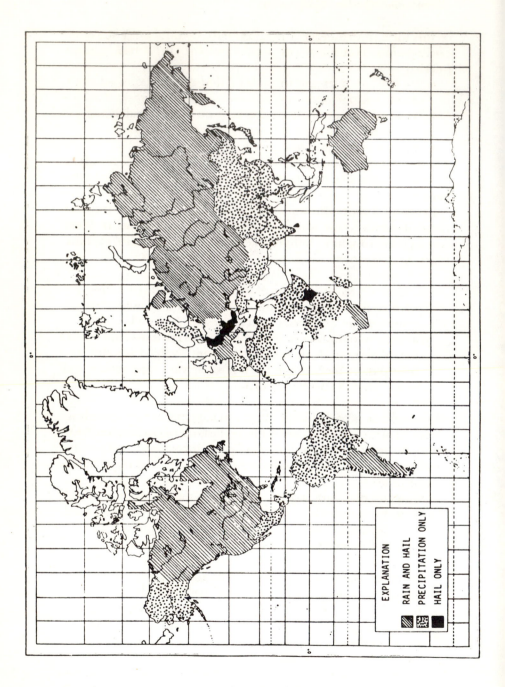

EXPLANATION

RAIN AND HAIL

PRECIPITATION ONLY

HAIL ONLY

rivers such as the Congo, Niger, and Nile, could provide a positive feedback mechanism, which would purposefully modify the climate:

> More water—more vegetation—more evaporation and less runoff—more clouds and rain—more cooling of the monsoon air (by evaporation and increased cloudiness)—more vegetation.

This mechanism, he noted, would shift the arable land limit northward and "presumably also the Tropical front itself because of increased production of cool monsoon air." As one can see, this plan somehow involves the controversial and as yet not well understood linkages of vegetation, irrigation, and cloud seeding on precipitation modification.

Concluding Remarks

During a recent study on the value of a potential "technological fix," i.e., the availability of a long range forecast for the West African Sahel, it became quite clear that many of the perennial problems confronting the inhabitants of the Sahel and their governments are not directly related to drought (Glantz, 1977). The drought, however, did highlight as well as exacerbate these problems. In a similar way, it might be argued that technological fixes, like climate and weather modification schemes, may raise expectations falsely and mask the underlying problems, as well as create new ones. On this point with respect to Soviet hydroengineering schemes, Soviet geographer Innokenti Gerasimov (1971) has written that

> [hydroengineering] solutions to some important economic problems have led to serious consequences that were not anticipated. The full extent of these unforeseen consequences has not yet been revealed, but what is

Figure 12.9. Nations in which weather (precipitation and hail) modification projects (experimental or nonexperimental) have occurred since 1945.

already known should undergo comprehensive scientific and economic analysis. This is necessary in order to forecast the impact of future hydroengineering schemes.

It can be argued that there now exists much information which, if used, could lead to an improvement in the quality of life of inhabitants in the marginal lands in Africa. With respect to the Sahel, it is widely acknowledged that there have been, for example, too many cattle grazing the Sahelian rangelands, that additional boreholes in the region have often exacerbated the deterioration of the ranges, and that the bureaucracies in many of these states have been too compartmentalized to deal effectively with interdisciplinary problems. Productivity could be increased within these countries, given improvements in these problem areas. One need not await a technological breakthrough in order to make needed changes. While the search for technological fixes for increasing the utility of marginal lands continues, human mismanagement of the relatively more productive lands is in the long run removing potentially good land from use.

Weather and climate modification *may* prove to be a long-term solution for the development and reclamation of potentially arable land in the Sahara and Sahel as well as for other regions in Africa. At this time, however, scientific research in the way of feasibility studies and experimentation has, for many of the proposed schemes, not yielded enough information about the impact or the utility of such schemes to make them operational (WMO, 1974). Yet solutions to perennial problems, caused in part by the variability of climate in the region, are needed now. Rather than awaiting the development of a new technological fix, it is strongly suggested that attention be turned to the improvement of poor land management practices used by the inhabitants of the area. This interim solution would seek to arrest the current deterioration of arable and grazing lands, thereby making them available for continued use in the future. It would be easier and less costly to stop the deterioration before the land is lost to desertification and erosion, for example, than it would be to rejuvenate the land once it has deteriorated beyond a critical level.

It has been generally accepted that human inputs are a major factor in the process of desertification. This is not a recent view. Twentieth-century scientists and ecologists have been trying to get this viewpoint across to political decision makers for a long time. E. P. Stebbing, for example, wrote several articles in the 1930s on the nature of human impact along the southern edge of the Sahara using such provocative titles as "The Encroaching Sahara . . . " (1935), "The Threat of the Sahara" (1937), and "Man-Made Desert in Africa" (1938).

Different governments have been aware of the negative impact that man has had on his environment, especially in semiarid zones. Historical evidence, however, tends to indicate that individuals, as well as governments, do not necessarily learn from past mistakes either directly or by analogy. Many of the proposals suggested in response to the drought crisis in the Sahel today are similar to those suggested 60 years ago, for example, in South Africa. The South African Senate report presented earlier, urging that attention be given to such destructive land management practices as erosion, overgrazing, misuse of water, grass burning, and deforestation, was issued in 1922. A Chinese comment on desertification was written in 1931 (Chu Co-ching, 1954), and one on Rhodesia in 1967 (West; Floyd, 1972). An awareness of the problems faced by governments in semiarid zones is, therefore, geographically as well as chronologically widely distributed.

Awareness of the problems, however, is not necessarily translated into action that will rectify those problems. The complexity of the problem was recently suggested by an official from a major donor country who had interviewed a Fulani herder in northern Upper Volta in the spring of 1973. He reported that the farmer, asked how he had been affected by recent drought, said he had had 100 head of cattle and had lost 50. The farmer continued, "Next time I will have 200," implying that by starting with twice as many he would save the 100 cattle that he wants. Yet the land's carrying capacity is such that he will still have only 50 cattle, but his loss will have been much greater.

A misleading feature of drought is that, following such a natural disaster, the population (human and livestock) is greatly reduced because of drought-related deaths or migrations. The land's carrying capacity will then be relatively more in balance with the drought-reduced populations that the land will have to support. What will have changed for the worse, however, will be the rejuvenative capacity of the land. One must ask, as Stebbing did in 1935, "How long before the desert supervenes?"

References

ALESCO/UNEP, 1975. Re-greening of Arab deserts: a pre-programming study. Cairo: Arab League Educational, Cultural and Scientific Organization (ALESCO)—United Nations Environment Program (UNEP) Project #0206-74-002, April.

Alusa, A., 1974. Weather modification activities in East Africa: A review. In *Preprints of the International Tropical Meteorology Meeting*. Nairobi: 31 January–7 February, pp. 302-307.

Ambroggi, R. P., 1966. Water under the Sahara. *Scientific American*, 214(5): 21.

Anon., 1976. African shortages disrupt A.I.D. plans. Denver *Rocky Mountain News*, 14 June.

Anon., 1975. Can you lead an iceberg to the desert? *Christian Science Monitor*, 21 November.

Anon., 1975. *New York Times*, 16 February, p. 11.

Anon., 1925. Drought, its prevention and cure. *The Meteorological Magazine*, December, p. 264.

Atlas, D., 1976. The paradox of hail suppression. National Center for Atmospheric Research mimeo., 17 May.

Bergeron, T., 1970. *Cloud Physics Research and the Future Fresh-Water Supply of the World*. Uppsala: University of Uppsala Meteorological Institution Report no. 19, p. 10.

Black, J. F., 1970. Asphalt island concept of weather modification. Linden, N.J.: ESSO [now EXXON] memorandum, 4 June.

Black, J. F., and A. H. Popkin, 1967. New roles for asphalt in controlling man's environment. Paper presented at the National Petroleum Refiners Association Annual Meeting, April 1967, at San Antonio, Texas, p. 6.

Black, J. F., and B. I. Tarmy, 1963. The use of asphalt coatings to increase rainfall. *J. of Applied Meteorology*, 2(5): 557.

Booker, R., 1973. *Project Rain—1973: A Final Report to the Republic of Niger (Niamey) and to Africare (Washington, D.C.)*. Norman, Okla.: Weather Science, Inc., November, p. 12.

Changnon, Jr., S. A., 1975. The paradox of planned weather modification. *Bulletin of the American Meteorological Society*, 56(1): 27-37.

Charney, J., et al., 1975. Drought in the Sahara: a biogeophysical feedback mechanism. *Science*, 187(4175): 434.

Chu Co-ching, 1954. Climatic change during historic times in

China. In *Collected Scientific Papers: Meteorology 1919-1949*. Peking: Academia Sinica, p. 272.

Colorado International Corporation (CIC), 1975. *Rainfall Augmentation in the Lualaba Watershed*. Boulder, Colorado, June.

Falkenmark, M., and G. Lindh, 1974. How can we cope with the water resource situation by the year 2015? *Ambio*, 3(3-4): 120.

Fletcher, J. O., 1971. Controlling the planet's climate. In *The Survival Equation*, R. Revelle, ed. Boston: Houghton Mifflin Co. Pp. 442-43.

Floyd, B. N., 1972. Land apportionment in Southern Rhodesia. In *People and Land in Africa: South of the Sahara*. New York: Oxford University Press. Pp. 225-26.

Gates, W. L., 1975. The January global climate simulated by a two-level general circulation model: a comparison with observation. *J. of the Atmospheric Sciences*, 32(3), March: 449.

Gerasimov, I. P., 1971. A Soviet plan for nature. In *Encountering the Environment*, A. Meyer, ed. New York: Van Nostrand Reinhold Co. P. 198.

Gibbs, W. J., 1975. Drought—its definition, delineation and effects. In *Drought*. Geneva: WMO Special Environment Report no. 5, p. 25.

Glantz, M. H., 1977. The value of a long range weather forecast for the West African Sahel. *Bulletin of the American Meteorological Society*, 58(2).

Glantz, M. H., 1975. The Sahelian drought: no victory for Western aid. *Africa Today*, April/June, pp. 57-61.

Grant, J., 1973. Some notes on drought in Africa and on control of weather. Paper presented at the University of London (SOAS) Symposium on Drought in Africa, 19-20 July, p. 1.

Gray, W., et al., 1974. Weather modification by carbon dust absorption of solar energy. In *Proceedings of the Fourth Conference on Weather Modification*, Ft. Lauderdale, Florida, 18-21 November, p. 195.

Gray, W. M., W. M. Frank, M. L. Covin, and C. A. Stokes, 1976. Weather modification by carbon dust absorption of solar energy, *Journal of Applied Meteorology* 15(4): 355-86.

Huff, F. A., and S. A. Changnon, Jr., 1973. Precipitation modification by major urban areas. *Bulletin of the American Meteorological Society*, 54(12): 1230.

Kelley, J., 1968. The Paraguana Peninsula as a potential site for weather modification. University Park: Pennsylvania State

University Department of Meteorology, January.

Kellogg, W. W., and S. H. Schneider, 1974. Climate stabilization: for better or for worse? *Science*, 186(4170): 1163-72.

Lattimore, O., 1938. The geographical factor in Mongol history. *Geographical Journal*, 91(1): 1.

Le Houérou, H. N., 1976. Greenbelts and desertization. Paper presented at the annual meeting of the American Association for the Advancement of Science (AAAS), Boston, Mass., 19 February: 3.

Le Houérou, H. N., 1973. FAO correspondence. Rome: UN/FAO reference PL 11/1; V/Ref/0.MR1210, October, p. 1.

MacDonald, J., 1962. The evaporation-precipitation fallacy. *Weather*, 17(5): 169.

Malkus, J. S., 1963. Tropical rain induced by a small natural heat source. *J. of Applied Meteorology*, 2(5): 547-56.

Mountjoy, A. B., and C. Embleton, 1967. *Africa: A New Geographical Survey*. New York: Praeger Publishers. P. 42.

National Academy of Sciences, 1973. *Weather and Climate Modification: Problems and Progress*. Washington, D.C.: U.S. Government Printing Office.

Otterman, J., 1977. Anthropogenic impact on the albedo of the earth. *Climatic Change*, 1(2).

Otterman, J., 1974. Baring high albedo soils by overgrazing: a hypothesized desertification mechanism. *Science*, 186(4163): 531-33.

Parliament of South Africa (the Senate), 1914. *Report from the Select Committee on Droughts, Rainfall and Soil Erosion*. 19 June, p. 5-6.

Peixoto, J., and M. Ali Kettani, 1973. The control of the water cycle. *Scientific American*, 228(4): 55, 60.

Purdom, J. F. W., 1973. Satellite imagery and mesoscale convection forecast problem. In *Preprints of the Eighth Conference on Severe Local Storms*. Boston: American Meteorological Society, pp. 244-51.

Rapp, R. R., and M. Warshaw, 1974. *Some Predicted Climatic Effects of a Simulated Sahara Lake*. Santa Monica, Ca.: RAND, March.

Rusin, N., and L. Flit, 1962. *Man Versus Nature*. Moscow: Peace Publishers. P. 111.

Schnell, R., 1974. Biogenic and inorganic sources for ice-nuclei in the drought-stricken areas of the Sahel—1974. *Report to the Directors of the Rockefeller Foundation*. New York, December, p. 3.

Schwarz, E. H. L., 1920. *The Kalahari or Thirstland Redemption.* Capetown: T. Maskew Miller, p. 103.

Seck, M., 1974. Artificial precipitation in Senegal. In *Preprints of the International Tropical Meteorology Meeting.* Nairobi: 31 January–7 February, pp. 111-13.

Shapley, D., 1974. Weather warfare: Pentagon concedes 7-year Vietnam effort. *Science,* 184(4141): 1059-61.

Stebbing, E. P., 1938. The man-made desert in Africa. *J. of the Royal African Society/Supplement,* 37(146), January: 13.

Stebbing, E. P., 1935. The Encroaching Sahara . . . *Geographical Journal,* 86(5): 509, 510.

Taljaard, J. J., 1976. Republic of South Africa Weather Bureau, Pretoria, personal communication, 27 January.

Thompson, G. A., 1912. A plan for converting the Sahara desert into a sea. *Scientific American,* 10 August, p. 114.

Tyson, P. D., T. G. J. Dyer, and M. N. Mametse, 1975. Secular changes in South African rainfall: 1880-1972. *Quarterly J. of the Royal Meteorological Society,* pp. 817-33.

UNEP, 1975. *Overviews in the Priority Subject Area: Land, Water and Desertification.* Nairobi: UNEP/PROG/2, February, p. 11.

Wade, N., 1974. The Sahelian drought: no victory for western aid. *Science,* 185(4147): 235.

West, O., 1967. The vegetation of Southern Matabeleland . . . (Rhodesia). In *Abstracts of Proceedings of the First Rhodesian Science Congress.* Bulawayo, Rhodesia: The Teachers' College, 16-20 May, p. 99.

Winstanley, D., 1975. The impact of regional climatic fluctuations on man: some global implications. In *Proceedings of the WMO/IAMAP Symposium on Long-Term Climatic Fluctuations,* Norwich, England: WMO Publication #421, August, pp. 479-491.

World Meteorological Organization (WMO), 1974. Report of the third session of the EC panel on weather modification/CAS working group on cloud physics and weather modification (Appendix F). Toronto, 28 October–2 November.

Index